Secure Sensor Cloud

Synthesis Lectures on Algorithms and Software in Engineering

Editor
Andreas Spanias, *Arizona State University*

Secure Sensor Cloud
Vimal Kumar, Amartya Sen, and Sanjay Madria
2018

Sensor Analysis for the Internet of Things
Michael Stanley and Jongmin Lee
2018

Virtual Design of an Audio Lifelogging System: Tools for IoT Systems
Brian Mears and Mohit Shah
2016

Despeckle Filtering for Ultrasound Imaging and Video, Volume II: Selected Applications, Second Edition
Christos P. Loizou and Constantinos S. Pattichis
2015

Despeckle Filtering for Ultrasound Imaging and Video, Volume I: Algorithms and Software, Second Edition
Christos P. Loizou and Constantinos S. Pattichis
2015

Latency and Distortion of Electromagnetic Trackers for Augmented Reality Systems
Henry Himberg and Yuichi Motai
2014

Bandwidth Extension of Speech Using Perceptual Criteria
Visar Berisha, Steven Sandoval, and Julie Liss
2013

Control Grid Motion Estimation for Efficient Application of Optical Flow
Christine M. Zwart and David H. Frakes
2013

Sparse Representations for Radar with MATLAB ™ Examples
Peter Knee
2012

Analysis of the MPEG-1 Layer III (MP3) Algorithm Using MATLAB
Jayaraman J. Thiagarajan and Andreas Spanias
2011

Theory and Applications of Gaussian Quadrature Methods
Narayan Kovvali
2011

Algorithms and Software for Predictive and Perceptual Modeling of Speech
Venkatraman Atti
2011

Adaptive High-Resolution Sensor Waveform Design for Tracking
Ioannis Kyriakides, Darryl Morrell, and Antonia Papandreou-Suppappola
2010

MATLAB™ Software for the Code Excited Linear Prediction Algorithm: The Federal
Standard-1016
Karthikeyan N. Ramamurthy and Andreas S. Spanias
2010

OFDM Systems for Wireless Communications
Adarsh B. Narasimhamurthy, Mahesh K. Banavar, and Cihan Tepedelenliouglu
2010

Advances in Modern Blind Signal Separation Algorithms: Theory and Applications
Kostas Kokkinakis and Philipos C. Loizou
2010

Advances in Waveform-Agile Sensing for Tracking
Sandeep Prasad Sira, Antonia Papandreou-Suppappola, and Darryl Morrell
2008

Despeckle Filtering Algorithms and Software for Ultrasound Imaging
Christos P. Loizou and Constantinos S. Pattichis
2008

Secure Sensor Cloud

Vimal Kumar, Amartya Sen, and Sanjay Madria

ISBN: 978-3-031-00399-8   paperback
ISBN: 978-3-031-01527-4   ebook
ISBN: 978-3-031-00015-7   hardcover

DOI 10.1007/978-3-031-01527-4

A Publication in the Springer series
SYNTHESIS LECTURES ON ALGORITHMS AND SOFTWARE IN ENGINEERING

Lecture #18
Series Editor: Andreas Spanias, *Arizona State University*
Series ISSN
Print 1938-1727   Electronic 1938-1735

Secure Sensor Cloud

Vimal Kumar
University of Waikato, New Zealand

Amartya Sen
Oakland University, USA

Sanjay Madria
Missouri University of Science and Technology, USA

SYNTHESIS LECTURES ON ALGORITHMS AND SOFTWARE IN ENGINEERING #18

ABSTRACT

The sensor cloud is a new model of computing paradigm for Wireless Sensor Networks (WSNs), which facilitates resource sharing and provides a platform to integrate different sensor networks where multiple users can build their own sensing applications at the same time. It enables a multi-user on-demand sensory system, where computing, sensing, and wireless network resources are shared among applications. Therefore, it has inherent challenges for providing security and privacy across the sensor cloud infrastructure. With the integration of WSNs with different ownerships, and users running a variety of applications including their own code, there is a need for a risk assessment mechanism to estimate the likelihood and impact of attacks on the life of the network. The data being generated by the wireless sensors in a sensor cloud need to be protected against adversaries, which may be outsiders as well as insiders. Similarly, the code disseminated to the sensors within the sensor cloud needs to be protected against inside and outside adversaries. Moreover, since the wireless sensors cannot support complex and energy-intensive measures, the lightweight schemes for integrity, security, and privacy of the data have to be redesigned.

The book starts with the motivation and architecture discussion of a sensor cloud. Due to the integration of multiple WSNs running user-owned applications and code, the possibility of attacks is more likely. Thus, next, we discuss a risk assessment mechanism to estimate the likelihood and impact of attacks on these WSNs in a sensor cloud using a framework that allows the security administrator to better understand the threats present and take necessary actions. Then, we discuss integrity and privacy preserving data aggregation in a sensor cloud as it becomes harder to protect data in this environment. Integrity of data can be compromised as it becomes easier for an attacker to inject false data in a sensor cloud, and due to hop by hop nature, privacy of data could be leaked as well. Next, the book discusses a fine-grained access control scheme which works on the secure aggregated data in a sensor cloud. This scheme uses Attribute Based Encryption (ABE) to achieve the objective. Furthermore, to securely and efficiently disseminate application code in sensor cloud, we present a secure code dissemination algorithm which first reduces the amount of code to be transmitted from the base station to the sensor nodes. It then uses Symmetric Proxy Re-encryption along with Bloom filters and Hash-based Message Authentication Code (HMACs) to protect the code against eavesdropping and false code injection attacks.

KEYWORDS

sensor cloud, risk assessment, IoT, encryption, cyber-physical system, access control, secure code, secure data aggregation, privacy

Contents

Preface

This book is designed to learn the secure sensor cloud computing paradigm which the authors have introduced, designed, and developed over the last eight years for supporting many different on-demand sensing applications. The contents in this book are, therefore, a combination of novel and useful research results, as well as well-tested schemes and protocols designed and developed by the authors. The book is self-contained as it discusses the preliminaries required to understand the sensor cloud enviornment including background on security and privacy schemes to understand the designed algorithms. We have also reviewed other work in the sensor cloud domain and related technology for completeness.

The traditional model of sensor computing with wireless sensors/devices imposes restrictions on how efficiently these devices can be used due to resource constraints at the device level as well as due to wireless constraints. This book discusses our sensor cloud architecture, which enables different wireless sensor networks, spread in a huge geographical area but owned by others to connect together and accessed on an on-demand basis by multiple users at the same time. Some applications of sensor cloud monitoring may include monitoring patients and doctors inside or outside a hospital, for monitoring traffic control, celestial navigation, emergency alarms, and for surveillance, exploration of enemy forces, war assessment, and for weather conditions and disasters such as tsunamis and earthquakes.

The book describes how virtual sensors assist in creating a multi-user environment on top of resource-constrained physical wireless sensors and can help in supporting multiple applications on an on-demand basis. Sensor nodes are susceptible to attacks including node capturing and compromising. Wireless communications can be eavesdropped, captured, or tampered. The infrastructure of a sensor cloud can be misused by malicious users. The virtual network topology can be very different from heterogeneous physical topologies underneath, which can complicate security. Most security best-practices cannot be used here due to the limitation in communication and computation capability both at the network and at the device level. Thus, this book first provides a series of well-researched schemes for risk assessment in sensor cloud applications. It then provides solutions for secure data aggregation, privacy preserving data integrity, and secure code dissemination in a sensor cloud.

The authors believe that this book will serve as a foundation for building secure sensor cloud applications in many different domains such as cyber-physical systems, smart-city, health-monitoring, disaster management, etc. In addition, it will serve as a natural transition to building infrastructures to support secure Internet of Things (IoT) applications.

This book is a good resource for a senior undergraduate or graduate course in a secure sensor cloud or secure sensor network area. The book will also help researchers and industry practicioners working in general areas of security, sensor networks, IoTs, and cloud computing.

Vimal Kumar, Amartya Sen, and Sanjay Madria
October 2018

Acknowledgments

The authors acknowledge the research grant support from the National Institute of Standards (NIST), Army Research Lab (ARL), and Army Research Office (ARO) for the secure sensor cloud research, most of which have been documented here in this book.

The authors would also like to thank Andreas Spanias from Arizona State University for inviting us to write this book.

Finally, we would also like to thank the Morgan & Claypool staff and editors for working with us to complete this manuscript.

Vimal Kumar, Amartya Sen, and Sanjay Madria
October 2018

CHAPTER 1

Introduction

The sensor cloud [19] is a new paradigm of computing for Wireless Sensor Networks (WSNs). It integrates multiple wireless sensor networks to share resources in a layered architecture and provide on-demand sensing as a service to different users at the same time. It decouples the users and the providers (owners) of sensing as a service in a sensor cloud network. The sensor cloud lowers the barrier of use of WSNs and makes it easier for users to create, deploy, and use sensing applications on-demand with embedded sensors and enables cross-compatibility of WSNs that use different platforms and technologies. Many different applications, such as environmental monitoring, disaster management, battlefield monitoring, and cyber-physical applications, can use sensor cloud infrastructure. In many of these applications, an integral part of building a successful WSN application is security. Although a sensor cloud makes it easier to deploy applications over multiple WSNs, it introduces new security challenges due to user-centric on-demand services. In this introductory chapter, we will discuss the security challenges that need to be tackled in order to build a secure sensor cloud infrastructure. We will start by taking a look at the wireless sensing devices and WSNs before discussing the concept of a sensor cloud in detail. We will then provide the aforementioned security challenges in building a sensor cloud and finally discuss four specific challenges that are integral to building a secure sensor cloud.

1.1   WIRELESS SENSING DEVICES AND WIRELESS SENSOR NETWORKS

Wireless sensors are small but versatile devices that can be equipped with various types of environmental, physiological, biological, physical, or chemical sensors. In addition to their sensing capabilities, wireless sensors also have a small amount of on-board memory, storage, computational power, and communication capability. They can be used as compact sensing and computational devices either as individual nodes or in the form of a distributed wireless network that not only senses data but also processes it before communicating it to the user. Wireless sensors are also referred to as motes. There are many different types of motes available in the market today from various manufacturers and they differ in the capabilities that they offer. One of the important aspects of a wireless sensor/mote is its physical size. The vision of projects such as Smart Dust and Neural Dust is to eventually have motes with micrometer or even nanometer dimensions that are smaller than specs of dust and can be deployed completely unobtrusively. While there is a substantial amount of work in progress [7, 8] focusing on reducing the size of motes, the currently available off-the-shelf motes are considerably larger in size. Two of the widely used motes in WSN research are the Mica2 and the TelosB motes from Memsic which are both only slightly larger than 50×30 mm^2. Figure 1.1 shows Mica2 and the TelosB motes as an example of the sensor hardware that can generally be found on current off-the-shelf motes.

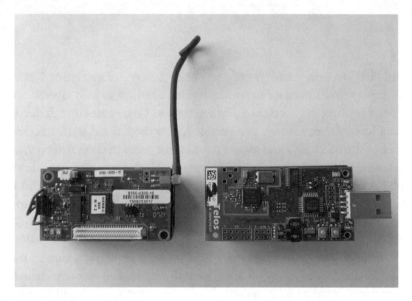

Figure 1.1: Mica2 and TelosB motes.

As mentioned above, wireless sensors can be used as individual sensing nodes but they can also be programmed to work in a distributed network environment. The on-board memory, microprocessor, and storage allows a wireless sensor to be programmed to do a specific task and the radio allows them to create a network with other wireless sensors. Such a network called a WSN covers a larger geographical sensing area than individual nodes. Wireless sensors can be programmed to self-organize in WSNs with any topology that fits the requirements to perform various sensing and communication tasks. The small size of these devices and their low cost makes them ideal to be deployed quickly and inconspicuously in places that may be hazardous for human beings or where setting up networking infrastructure may not be feasible for any other reason. WSNs, therefore, have applications in many areas such as battlefields, environmental monitoring, habitat monitoring, structural monitoring, healthcare, heavy industry etc. Some other examples of such an application is in battlefields where sensing may be required but setting up sensing and networking infrastructure may not be possible. Another application of wireless sensors is in measuring volcano-seismic activity [10], where the terrain itself is hazardous and ever-changing, and setting up permanent structures is not feasible. Readers interested in sensors and WSNs can further read [125, 126], and [127].

The small size of wireless sensors is very useful and is their prime feature, however, it gives rise to a number of processing constraints. The small size limits the amount of resources available on board. This in turn means that the processor, RAM, flash memory, sensors, communication range, and power are all very limited. Table 1.1 shows the meager specifications of the two commonly used motes; Mica2 and TelosB. In addition to the constraints, low-cost motes are generally not built keeping security in mind. This concoction of usefulness, constraints, and lack of security has given rise to a large body of research work focused on making the best possible use of this technology.

Table 1.1: Specifications of Mica2 and TelosB motes

Mote/Specifications	Mica2	TelosB
Processor	ATmega 128L	MSP430
RAM	4 KB	10 KB
Flash Memory	128 KB	48 KB
Data Bus Width	8-bit	16-bit
Data Rate	19.2 kbps	250 kbps
Clock Frequency (max.)	8 MHz	8 MHz
Current Draw (active/idle)	8 mA/15 µA	1.8 mA/5.1 µA
Transmission Frequency	868/916 MHz	2.4–2.483 GHz
Power	2 x AA batteries	2 x AA batteries

1.2   SENSOR CLOUD

WSNs clearly have huge potential and in the last two decades there has been a tremendous amount of research on various aspects of WSNs related to programming them, deploying them, securing them, and using them in various applications. However, as often happens with new and exciting technologies, the development happens rapidly all over the world resulting in growth that is not synchronized with growth in other parts of the world. This has led to the development of various technologies for the functioning of WSNs that are not compatible with each other. This gives rise to situations where two WSNs that want to collaborate cannot easily communicate with each other if they are not using the same technologies. This lack of compatibility has conspired to limit the proliferation of WSNs. Another factor that has contributed to preventing us from realizing the full potential of this technology is the development model that is currently in use. In this development model, which we call the traditional model, a network user is tightly coupled with the network owner. In this model, a user who wants to use a WSN will often need to own a WSN, program it, deploy it, and maintain it. The initial capital cost and the associated maintenance cost present a huge barrier to entry for small- to medium-scale organizations. Many WSN applications only require sensor data for a short period of time, or periodically with large intervals. This means that wireless sensors in a WSN are idle for most of their lifetime.

For medium- to large-scale organizations that incur the ownership and maintenance costs, this results in a low return to investment ratio.

The sensor cloud model resolves three main challenges described above. It first makes it easy for incompatible networks to work together for a common objective. Second, it lowers the barrier to entry for new users by providing sensing as a service. Third, it increases the WSN throughput and, therefore, profitability for WSN owners. The merits of a collaborative sensor cloud-like approach described in [13] was one of the earliest works in this area. The authors called their collaborative sensor network a "Virtual Sensor Network" (VSN). The VSN thus described was made up of multiple sensor networks that could work jointly if needed and share resources with each other. The VSN approach divided the sensors into two groups; one group that took part in data collection and one which communicated control messages. While the data collection nodes performed the sensing activities, the maintenance nodes performed other subsidiary tasks, such as communication of control messages, tasks associated with addition and deletion of nodes from the network, communication of broadcast messages, etc. This helped in reducing the burden over the resource constrained nodes that were performing the critical task of sensing.

Early approaches, such as [13] and later [14], were limited and were focused on creating an overlay VSN over multiple WSNs. Such approaches did not delve much into utilizing the potential of the middleware to open the network to multiple users. In [15], the authors presented their vision of a sensor cloud-like system. This system is a three-layered network, with sensors at the bottom layer that produce the actual data, a brokering network in the middle that filters unwanted and unimportant data and provides a publish-subscribe architecture and cloud at the highest layer providing computational power. In this work, smartphones are considered as the sensors at the bottom layer. Another similar system called IoTCloud is proposed in [16], which attempts to create a framework for real-time processing of data from smart devices by using a middleware in the cloud.

In general, a sensor cloud can be defined as a heterogeneous multi-user, multi-owner computing environment that brings various WSNs together, to work as one system for providing sensing as a service to users. The core of a sensor cloud is a software stack, i.e., middleware that can reside in the cloud. As shown in Figure 1.2, individual WSNs with sensor cloud membership are linked to the stack and provide data to it. This data could be continuous streaming data or may be provided on-demand. On the other end, users also connect to the stack and pay to receive the sensor data as a service.

1.2.1 SENSOR CLOUD LAYERED ARCHITECTURE

The software stack consists of three layers—the sensor-centric layer, the middleware, and the client-centric layer [19] (as shown in Figure 1.3).

The sensor-centric layer is the lowest layer in the stack and it directly communicates with wireless sensors. This layer is responsible for WSN registration, WSN maintenance, and data

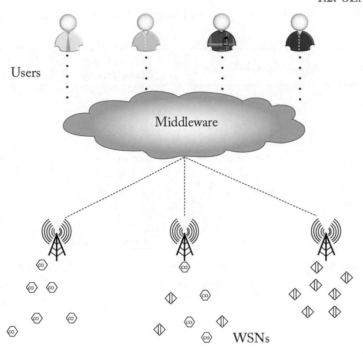

Figure 1.2: A birds-eye view of a sensor cloud.

collection. WSN owners register their network with the sensor cloud through this layer. The registration process creates a catalog of the data provided by the WSN and the associated QoS constraints and results in a contract between the sensor cloud and the WSN owner. After this, the owners are responsible for maintaining the network and providing data and the sensor network platform to sensor cloud according to the contract. The next layer in the stack is the middleware, which connects the data collected from the sensor-centric layer to the users at the client-centric layer. The middleware layer performs a number of important functions such as provisioning sensors to user requests, managing and overseeing sensor and data usage for billing and managing the sensor life-cycle. The top layer is the one where users interact with the sensor cloud referred as the user-centric layer. This layer provides an interface to the user and handles session and membership. This layer also uses the information provided by WSN owners to provide service catalogs to the users and customizing the service-layer agreements (SLAs).

1.2.2 VIRTUAL SENSORS

Like any cloud platform, the key idea behind sensor cloud is the virtualization of resources. The resources in a sensor cloud being the wireless sensory devices; energy, bandwidth, computing etc. Virtualized wireless sensors are generally known as virtual sensors as opposed to the physical

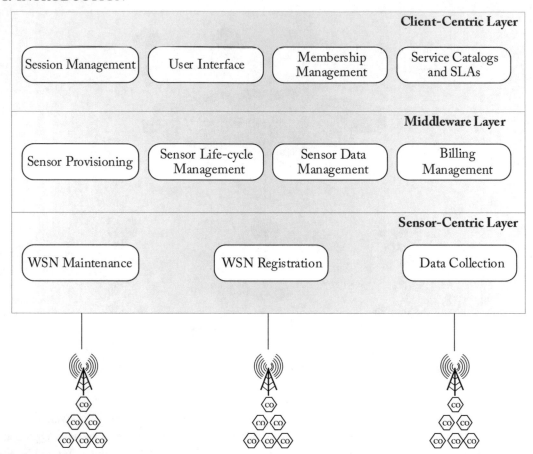

Figure 1.3: Sensor cloud layered architecture.

wireless sensors. Multiple virtual sensors can be executed on a single physical sensors. Virtual sensors can be used to improve resource utilization in WSNs. They can also be used to create complex sensors by combining multiple virtual sensors and computing capability.

The following two main approaches have been used in sensor clouds for achieving virtualization of wireless sensors:

- true virtualization through Virtual Machines (VMs) and

- pseudo virtualization through middleware.

True Virtualization

True virtualization systems run a thin layer of software on the mote hardware. WSN application software is then added on top of the thin software layer. Maté [18] was the first attempt at virtualizing a very low spec sensor mote. Maté is a bytecode interpreter written in TinyOS, which runs on wireless sensors. This bytecode interpreter receives functionality in the form of code in packets which can be a maximum of 24 bytes in size. When the base station wants the wireless sensor to change its functionality, it sends the code for the functionality in a packet. The wireless sensor receives this code, where Maté interprets the code and responds to the instructions in the code. This is an excellent approach when the program logic is simple but for complicated programs, especially those involving complex security algorithms, the 24-byte limit makes this approach infeasible. After Maté, many other systems to virtualize wireless sensors were proposed. Notable among them were VM* [17], MANTIS [11], DAViM [20], and Darjeeling [12]. Every true virtualization system, however, suffers from the same problem as Maté. The virtualization (hypervisor) layer consumes valuable resources on an already resource constrained device, leaving precious little for the actual application to run. This eventually means that only simple and non-sophisticated programs can be run on the wireless sensors with true virtualization.

Pseudo Virtualization

Pseudo virtualization, as opposed to true virtualization, takes place in the middleware. A virtual sensor with pseudo virtualization is a representation of the physical sensor in the cloud. This virtual sensor contains metadata about the physical sensor it represents and holds the data produced by the physical sensor. If needed, the virtual sensor may also have data processing code to respond to complex queries from the users. This virtual representation of a wireless sensor allows the sensor cloud to use the sensor and its data in a variety of ways, while allowing for location and distribution transparency [19].

1.2.3   SENSOR CLOUD DELIVERY MODELS

Sensor clouds offer two delivery models to users: **Infrastructure as a Service (IaaS)** and **Data as a Service (DaaS)**. IaaS provides the wireless sensor hardware directly for the user to use. In this model, the user interacts with the hardware through sensor cloud but is free to implement applications and code of their choice. The sensor cloud provides the networking environment and the ability to reprogram the sensors with the user supplied code.

In DaaS, the sensor cloud makes use of pseudo-virtualized virtual sensors to delivers data streams to the users according to their subscriptions. In this model, the sensor cloud assigns virtual sensors to users according to their requirements. Under the hood, however, multiple virtual sensors belonging to various users may be mapped to a single physical sensor. In other instances, a single virtual sensor may be mapped to data streams from multiple physical sensors, fusing the data and outputting a single data stream.

1.3   SECURE SENSOR CLOUD

The sensor cloud computing platform introduces numerous advantages over using traditional WSNs. The users who subscribe to the services of the sensor cloud do not need to own or maintain their sensor networks. Multiple users requesting various multi-sensing activities for multiple applications can simultaneously rent the services offered by large-scale heterogeneous WSNs according to a pay-as-you-use model. This is facilitated by projecting the physical sensors over multiple virtual sensor networks and vice versa, as mentioned in the previous sections. The implementation of such a model enables resource sharing and efficiently allows for scaling up or down according to the demands of the users. The sensor cloud abstracts different physical platforms, giving its users the view of a dedicated homogeneous network servicing their application requests. This in turn greatly enhances a user's quality of experiences and allows them to focus on the functionality of their applications instead of management or maintenance of physical devices.

The sensor cloud paradigm clearly introduces many benefits over the utilization of traditional WSNs. However, the new paradigm also gives rise to new security challenges. The apprehensions about security primarily stems from the following factors:

- different ownership entities for physical WSNs,

- integration with a cloud platform and resource sharing, and

- multiple users subscribing to the services of the sensor cloud with different type of applications.

The different assets associated with the sensor cloud platform are summarized in Figure 1.4. Several vulnerabilities and threats are linked with these assets of sensor cloud platform which needs to be addressed in order to securely facilitate the benefits introduced by the sensor cloud paradigm to its users.

Cloud Platform: supports the middleware stack of the sensor cloud. One of its primary functionalities is hosting the virtual sensor networks which allow users to interact with the set of physical sensor nodes serving their application by propagating instructions and codes to be run on them, and obtain the data sensed and sent back by the physical devices. In doing so, this platform also enables resource sharing. As a result of these services, there can be two notable threats associated with this asset. First, there is a possibility of Malware Injection by allowing user applications to propagate instructions or codes which will eventually be executed on the physical sensors and in turn might compromise them. Second, resource sharing via virtual sensor networks might be used by an adversary to monitor the traffic and usage behavior of legitimate users if proper isolation is not provided. Further, resource monitoring also presents the risk of resource depletion which can be caused by an adversary by demanding more resources during the peak usage time of legitimate users.

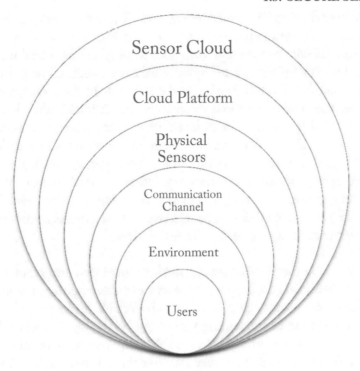

Figure 1.4: Sensor cloud assets.

Physical Sensors: are the infrastructure backbone of sensor cloud computing platform. These sensor nodes are heterogeneous in nature and have different ownership entities. This results in different WSNs (comprising of these heterogeneous sensors nodes) to have varying networking and security policies. Hence, the aspect of implementing and maintaining universal security measures becomes a non-trivial task. Furthermore, as depicted in Table 1.1, the sensor nodes are resource constrained in terms of their operating energy, storage memory, and processing powers. This renders the incorporation of the traditional security measures which are used in the domain of wired networks like public key encryption schemes, inefficient. Additionally, these sensor nodes are deployed in ad hoc fashion over vast geographical regions and are left unattended over long time periods. These factors make the sensor nodes vulnerable to attacks such as node capture, node malfunction, or outage either due to running out of energy or hostile environments. A feature unique to sensor clouds is that data generated by some sensors may be communicated to the cloud through other sensors owned by a different entity. This introduces a new attack vector in sensor cloud computing.

Communication Channel: for the physical sensor nodes is wireless in nature. In contrast to wired networks, wireless communication does not require the setup of a pre-defined infrastructure in terms of laying cables and the availability of ports. This considerably reduces the cost with respect to infrastructure downtime and monetary expenses. Further, the wireless communication medium can also be used with relative ease (compared to wired communications) in hostile environments like war zones or regions affected by natural disasters such as hurricanes, floods, and so forth. However, the notable disadvantage lies in the fact that the data transfer speed of wireless communications are much less than wired communication. Additionally, wireless communications can also be unreliable at times due to interference from the environment or obstruction due to presence of physical objects in the line of communication. Wireless communication is also typically more vulnerable than its counter part. Unencrypted communication presents the risk of eavesdropping or snooping by an adversary. Other forms of risks are also present in the form of data packet capture, tampering, and false data packet injection. Finally, it is relatively easier for an adversary to execute denial of service attacks in a wireless communication medium by simply jamming the radio signals of the physical devices.

Environment: in which the physical sensor nodes are deployed also plays an important role in determining the likelihood and impact of threats and vulnerabilities on sensor cloud infrastructure. For example, the security risks confronted by WSNs deployed across a war zone or a disaster prone region will not be the same as that of WSNs deployed in a hospital or industrial area. The nature of the environment in which the WSNs are present determines the ease of execution of security risks confronted by another asset like the communication channel or physical sensors. To elaborate this, consider a scenario in which the WSNs are easily discoverable by adversaries. This will essentially increase the likelihood of risks such as node capture or denial of service. Hence, it is essential to evaluate the deployment regions of the physical infrastructure, and develop counter measures that addresses the security risks arising due to them.

Users: lie at the core of all the sensor cloud assets. The exploitation of any security risks present in any of the aforementioned assets will ultimately affect the users who subscribe to the services of the sensor cloud. However, the users themselves are prone to different security risks and threats. Adversaries can pose as legitimate users and can execute applications which host malware infected codes. Legitimate users are also vulnerable to social engineering attacks which may leak their identities and be used by adversaries in an unauthorized fashion. The task of addressing the security risks confronted by the users is more challenging than the other assets since some of the activities performed by the user falls beyond the trust boundaries of the sensor cloud platform, e.g., the devices that a user operates in order to interact with sensor cloud platform. The security of such devices is beyond the scope sensor cloud platform and can be exploited by adversaries which in turn can compromise sensor cloud.

The traditional security measures that one might incorporate to address most of these security risks and threats are not applicable in the domain of sensor cloud. The primary reason behind this is that the physical sensors are resource constrained, both in terms of computation and communication capabilities. As such, to identify the various threats confronted by the aforementioned assets and design effective and efficient security measures, the security of sensor cloud is addressed in the following three phases:

- secure pre-processing,

- secure processing, and

- secure runtime.

Phase I: Secure Pre-Processing

The sensor cloud is a unique paradigm which allows its users to rent services provided by its heterogeneous WSNs. In doing so, the users through their applications have the capabilities to interact directly with sensor cloud network and infrastructure. This is enabled by allowing users to pass different commands to the physical sensors which can range from entire code blocks, script functions to simple command line arguments. In terms of security risks, this may lead to execution of the different malware attacks if the software propagated to the physical sensors from the users is not sufficiently assessed and restricted. The physical sensors themselves are vulnerable to a plethora of security risks which can compromise the identity of different users they are serving or violate the secure functionality of the entire sensor cloud platform.

In the presence of such a high amount of cross-layer dependencies, incorporating and adhering to a set amount of pre-determined security policies is not going to guarantee the secure functionality of the sensor cloud platform. Especially, when the sensor cloud is unique with respect to its traditional counterparts like the WSNs or cloud computing platform. Hence, to design and incorporate effective and efficient security measures for sensor cloud, it is imperative to perform the task of threat modeling and risk analysis. This task can be captured (1) in the process of performing a system wide security risk assessment incorporating the cross-layer dependencies, (2) environment in which the physical infrastructure is deployed, (3) subjective assumption of the adversarial capabilities, and (4) effect of considered security measures in suppressing the identified security threats.

Risk Assessment in Sensor Cloud.   A sensor cloud consists of various heterogeneous WSNs may have different owners and run a wide variety of user applications on-demand in a wireless communication medium. Hence, they are susceptible to various security attacks. Thus, a need exists to formulate effective and efficient security measures that safeguard these applications impacted from attack in the sensor cloud. However, analyzing the impact of different attacks and their cause-consequence relationship is a prerequisite before security measures can be either developed or deployed. In this topic, a risk assessment framework for WSNs in a sensor cloud is presented that utilizes attack graphs. Bayesian networks are used to not only assess but also to quantitatively analyze the likelihood and impact of attacks on WSNs. The risk assessment framework will first review the impact of attacks on a WSN and estimate reasonable time frames that predict the degradation of WSN security parameters like confidentiality, integrity, and availability. Using the presented risk assessment framework will allow security administrators to better understand the threats present and take necessary actions against them.

Phase II: Secure Processing

The sensor cloud platform can have numerous users with varying objectives, different assets interconnected through cross-layer dependencies, and through all this its core mission is to facilitate secure services to its users. To achieve this, it should be equipped with various security measures to maintain the desired level of security in the presence of different real-world security risks. In the processing phase, two of the primary security concerns are as follows.

Identity and Access Management.   The sensor cloud platform has a distributed architecture which aids the provisioning of Business-to-Business collaborations. Different WSN owners register their networks with the sensor cloud and users rent the available sensing services for their application in an on-demand basis. In such a multi-user, multi-owner system, user access control is a significant problem. A system needs to exist that identifies a user and then authorizes it to only access data that was paid for by the user. While many user access control techniques exist, the unique characteristics of a sensor cloud make those techniques unsuitable to employ. An access control system for sensor cloud needs to take into account that data will be collected from resource constrained sensors and will generally arrive as a stream. This data may also be aggregated on the sensors. The system also needs to take into account the WSN owner's control over their devices and should allow for easy billing, management, and revocation of users.

Data Privacy.   The sensor cloud paradigm allows the creation of virtual sensor networks which can be projected onto the physical sensor and shared among different users. Data privacy is one of the vital concerns in the light of such resource sharing. The advent of data mining techniques allows adversaries with the capabilities to collect user data and create new information sinks by combining data from a wide variety of sources with relative ease. Therefore, the privacy breaches in sensor cloud-based services can be difficult to detect as the sensor cloud is provisioned for on-demand services. This implies not only ineffective defenses but also undue cost and wasted resources. In particular, a virtual sensor network can be abused to gather sensitive information or breaching user information privacy. This kind of risk is known as function creep [9]. The function creep occurs when an item, process, or procedure designed for one specific purpose end up serving another purpose. The function creep in the sensor cloud is more difficult to prevent. In addition to data privacy, there is a potential for an adversary to breach the meta-data, thereby revealing the identity of the physical sensors that are serving the users. This problem is similar to Yao's Millionaires' problem. The adversary may take advantage of the service negotiation to reveal the secrets or ownership of the physical WSNs. For example, physical sensors are different in various aspects of device specifications. Mica2 and TelosB sensors shown in Table 1.1 are significantly different in memory size, current draw, transmission frequency, and bandwidth. It is possible for an attacker to craft the sensing operation/mission in such a way that only one specific sensor can operate on it, thereby revealing which physical sensors are being used. The challenge is to balance the anonymity and the service availability so as to yield the optimal benefit of the sensor cloud services to the user as well as the provider.

Phase III: Secure Runtime

During the execution of the sensor cloud missions, one must ensure the security of its services, along with the security measures ascertained during the secure pre-processing and processing phases. The security risks that are present during the runtime phase are the compromise of the data packets transmitted by the physical sensors back to the user, capture of user code which is transmitted from the cloud platform to the physical sensors when they are provisioned, and compromise and exploitation of the security keys used by the physical infrastructure in order to communicate securely. These risks are addressed by the following security techniques which are adopted during the secure runtime phase of the sensor cloud platform.

Secure Data Aggregation.   Wireless communication for data transmission is one of the most energy consuming tasks performed by the physical sensors. In order to maximize the return of investment for the WSN owner, it is essential to optimize the network lifetime. This challenge is addressed by the usage of data aggregation techniques. Data aggregation techniques that are also secure are a pre-requisite in the presence of adversaries being able to physically compromise the WSNs in sensor cloud. Many secure and trust based data aggregation models assume firm network connectivity, where each sensor monitors the behavior of its neighbors. Such an assumption is challenging and to a certain degree prohibitive in the sensor cloud environment. Sensor cloud topology presents three challenges in the formulation of a secure data aggregation technique:

- a WSN owner may not allow sensors from a different network to monitor its sensors;

- a virtual sensor network for a user projecting to different physical sensors may be located far away from each other and may have no means of directly monitoring their virtual neighbors; and

- an individual WSN may be compromised completely by an adversary, which will render the "monitor your neighbor" approach useless.

In addition to this, when data is aggregated at each hop in a sensor network, it becomes harder to protect its integrity. A number of secure data aggregation algorithms have recently appeared for WSNs, very few of them, however, also address the issue of data integrity. In a secure data aggregation environment, it becomes easier for an attacker to inject false data. Even though data confidentiality and integrity are two contrasting objectives to achieve when data is aggregated, we suggest that both of these should be treated together.

Efficient and Secure Code Dissemination.   The WSNs in the sensor cloud platform are dynamically provisioned on a demand basis. When sensors are provisioned to users, new application code may need to be installed on the sensors. The inclusion of new users during the runtime of the sensor cloud will require the propagation of their application (and code) on the physical sensors which will be tasked to service them. As number of users in the system increase, the frequency of de-provisioning and provisioning of sensors increases. This in turn implies that application code on the sensors needs to be changed very often. When application code is transmitted to sensors, two things needs to be taken into consideration. In a sensor cloud, the application code is disseminated from the base station and reaches the target group of sensors through multi-hop communication over other wireless sensors. These wireless sensors over which the code travels may belong to different owners and may be provisioned to various users. First, forwarding packets (i.e., application code in this regard) too often consumes considerable amount of energy. The physical sensors are resources constrained in terms of the energy and available bandwidth. Therefore, the application code dissemination needs to be performed in an energy efficient fashion. Second, the physical sensor tasked to serve a user may belong to different WSNs. In a sensor cloud environment, different wireless sensors networks have different ownership entities. These WSNs are tasked to serve their internal users as well as external users (sensor cloud customers). Hence, a user's application code propagated to these physical sensors needs to be done in a confidential fashion along with preserving their integrity. To ensure the security of the data generated by the application code, it is critical to ensure the integrity and the confidentiality of the application code as it travels over the wireless sensors. Thus, we require a code dissemination algorithm that it is able to provide confidentiality and integrity of the code being transferred. This code dissemination algorithm needs also to be energy efficient.

Key Management.   Confidentiality in WSN missions is quite important. For example, the breach of confidentiality in military applications by an adversary may result in incorrect decision making (hostile deployment of troops in a war zone) leading to loss of human life. In sensor cloud service environment, confidentiality manifests in newer dimensions. Users of any sensor cloud will desire confidentiality of their tasks and communications, which means a desired level of the security needs to be provided to the users. Since multiple applications may run simultaneously on a WSN at the same time, information exchanged among sensors from one application must be isolated from another application. Additionally, networks of WSNs may have to collaborate for the same mission. An adversary compromising one WSN should not be able to compromise collaborating WSNs. All these situations are unique to sensor cloud-based infrastructures and need new solutions to address the challenges.

The overview of the security phases and the security challenges associated with them is summarized in Figure 1.5.

In the following chapters, the details of the implementation of the sensor cloud architecture is presented along with the discussions of the four prominent security challenges of *risk*

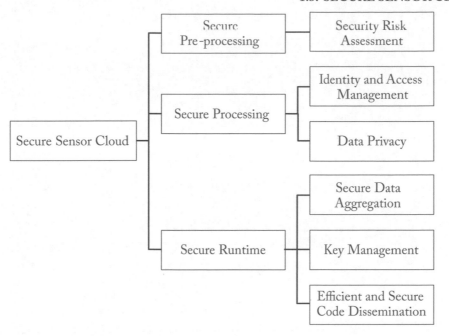

Figure 1.5: Security phases in a sensor cloud.

assessment, secure data aggregation, access control of data, and efficient secure code dissemination in a sensor cloud environment.

CHAPTER 2

Preliminaries

2.1 SECURITY RISK ASSESSMENT

In the absence of budgetary constraints, it is not an effective strategy to simply implement numerous random security measures. Since, without an understanding of the risks associated with the platforms which hosts an organization's application, developing and implementing security measures are futile leading to inefficient utilization of limited security resources. Therefore, addressing cyber security threats by formulating efficient and effective security measures is a multi-step process which requires performing security requirements analysis, risk assessment, risk management, and mitigation. These processes will also help organizations understand the benefits of their investment in cyber-security.

Security risk assessment lies at the crux of this multi-step process, enabling the incorporation of organization-wide assessment to determine the security threats, vulnerabilities, and their impact on network security parameters like confidentiality, integrity, and availability. It also helps in prioritizing the risks that needs to be addressed thereby playing an important role aiding the allocation of security measures constrained by an organization's budget. However, performing idealistic risk assessments for a large-scale organization can still be very expensive and may produce results which might require considerable domain knowledge in order to comprehend and generate useful insights. Therefore, the task of risk assessment is not trivial and requires careful consideration. In the following sections, we discuss some of the many facets of generic risk assessment methodologies.

2.1.1 RISK ASSESSMENT METHODOLOGIES

The process of security risk assessment is guided by identifying security requirements and thereafter performing extensive evaluations of an organization's Cyber security prospects. These tasks involve assessing the network infrastructure for security threats due to vulnerabilities that are present and the attacks that can exploit them. Risk assessment also enables the evaluation of another important organization asset—the people. Users of an application and an organization's employees are also susceptible to different threats which can result in the Cyber security incidents. These threats cannot be detected by running a vulnerability scanner tool on the network, further necessitating the need of performing risk assessment. At the end, risk assessment provides a detailed report on the threats confronted by an organization in different domains (network, people, security policies, etc.). This can be used in the risk management and mitigation process to either patch security vulnerabilities and prevent certain attacks or to pro-actively develop prioritized security incident response plans to minimize the effect of an exploit. Therefore, risk assessment acts as an estimation and validation tool for an organization by shedding light into security return on investment, get feedback on implemented security measures. This is also help to understand their cohesion with respect to best security practices, compliance and standards as laid down by regulatory bodies like NIST or ENISA.

Security Risk Assessment Process

The security risk assessment process is used to determine the strengths and weaknesses of an organization's systems, identifying and minimizing threats below a threshold which is acceptable as per the security requirements of the organization. An example of an organization's security requirements could be ensuring a certain level of assurance in confidentiality, integrity, and availability of their application and the data it processes. Risk assessment generally focuses on evaluating the likelihood of an undesired event (for, e.g., a data breach, or an unauthorized access) and the impact it will have on the exploitation of the system and the organization as a whole. Once this is evaluated, risk mitigation measures are designed and developed to minimize the likelihood and impact of the risks. In broader terms, risk assessment can be used to identify risks in different areas of an organization and not just related to Cyber security. For example, traditional risk assessment in the domain of information security and IT security can be used to evaluate systems and applications that support the functional services of an organization, its network and servers, physical security of the devices and premise, risks present (due) to employees of the organization. The process of risk assessment is normally (and must be) utilized during the conception of an IT service. Other than this, addition of new functionalities to an application/service; changes in the networking environment; change in technology (Software or Hardware updates) should also prompt the utilization of the risk assessment process.

The scope of attacks sustained by a WSN has been surveyed and discussed in [52, 53, 56]. The authors have assessed well-known sets of WSN attacks along with their countermeasures. They were, however, oblivious about the attack's impact on a network and efficiency of the countermeasures. In his survey on security issues of a WSN, Walter [38] established the parameters based on which security of a WSN is characterized. These parameters were confidentiality, integrity and availability. We are able to design the attack patterns from this information for our work and analyze the attacks on a broader perspective. Analysis of various attacks adopted by the adversary to exploit security parameters and ways in which they could be averted were also discussed in [38], although it did not address the likelihood of exploitation of an attack. Wood [57] and Xu [58] give exposition on the omnipresent denial of service (DoS) attack. DoS are not only hard to predict but also to counter. This helped us in understanding the nature of jamming attacks in WSNs. The absence of predictability and correlation with other attacks in case of DoS attack is a drawback on the security administrator's part. Karlof [59], Kannhavong [60], and Newsome [37] give an in-depth analysis on routing layer attacks and the Sybil attack, respectively. However, these attacks can be exploited by successful execution of attacks in different network layers. For this purpose we should identify the interdependencies between different feasible attacks. Mauw [61] and Phillips [50] demonstrated this kind of logical relationship via attack graphs or trees. Using the principles from the work of Lee [62], we were able to assess the risks to a network. But the drawback was that they were for a wired network scenario. Sheyner [49] discussed the various types of attack graph and models. This contributed immensely toward the development of the attack graph model for WSN. Gallon [63] devised the methods to quantitatively assess the attack nodes in the attack graph, although they were not meant for a WSN. National vulnerability database [44] established the vectors to calculate the severity ratings of vulnerabilities in a wired network. Using the same principles, we calculated the severity ratings for the attacks on WSN. Frigault [45] gave insight on implementing attack graphs as a Bayesian network. This gave us a better understanding about the adversary's capability, likelihood, and impact of attacks for various attack scenarios. Dantu [64] and Liu [65] analyzed attacks by assigning probability values to the attack graph nodes. Furthermore, using the concepts of Bayesian networks on these probability values, they calculated potential attack paths and modeled network vulnerabilities. Nevertheless, these computations were not for an attack scenario of a WSN and as such could not be applied to the proposed attack graphs for a WSN. Houmb [46] proposed the methodologies of risk level estimations using the exploitation frequency and impact of vulnerabilities in a wired network. We adopted these concepts to identify the metrics necessary to compute net threat level to the root node of our attack graph when it is represented as a Bayesian network. This gave a degree of diversification and uniqueness to the WSNs with respect to quantitatively analyzing our attack graphs and using the results to estimate maintenance period for the largely unattended WSNs.

Organizations must follow through with the creation of risk assessment policies which outlines a blueprint to guide the assessment process to be carried out. Such a blueprint consists of guidelines, establishing factors such as the following.

- When does an organization need to perform risk assessment?

- How often should it be repeated?

- What will be the scope of risk assessment (how comprehensive it is going to be)?

- Who will be in charge of carrying out the task (internal or third party)?

- What is expected from it (actionable insights)?

- Prioritizing risk levels (what sort of risks will be acceptable and which ones will be deemed critical).

- What methods will be used to perform risk assessment (qualitative, quantitative or hybrid)?

The risk assessment process, an exhaustive task, is initiated with the risk assessment policies document. It helps in defining the scope of the task and appointing personnels who will be responsible to carrying it out. Thereafter, risk assessment procedures are chosen (or developed) and a list of threats are identified which is followed up with identifying vulnerabilities. In this context, a *Threat* is defined as an unwanted event that may cause harm; *Vulnerability* is defined as a weakness which may provide a way for a threat to materialize; *Impact* is defined as the consequence(s) of a threat that has materialized. After vulnerability identification, security measures are determined and evaluated which might either help to mitigate, transfer (getting insurance policies), or avoid the threat and its impact. Thereafter, all the aforementioned information are used to estimate probability values which will depict the likelihood of occurrence of the threat in the presence of evaluated security measures. This is generally done either by using experts with domain knowledge, historical logs of threats, or statistical analysis. For prioritizing, the estimated probability can be further categorized as *Very likely*, *Isolated incidents*, *Rare*, *Very unlikely*, and *Almost impossible*. These categorizations are subjective and typically depends on the organization and their risk assessment policies and followed up with sensitivity analyses like Monte Carlo analyses.

Once probabilities are estimated, the damage is quantified in terms of the impact of the identified threats when they exploit the vulnerabilities. This followed up with risk level estimations which is defined by threat multiplied by its likelihood and impact probabilities. Although organizations like NIST have a scale for categorizing risk levels (for, e.g., 1.0: High, 0.5: Medium, 0.1: Low), it is a challenging task to be able to quantify and categorize risk levels based on monetary loss or loss of reputation. After risk level estimations, security measures are re-evaluated and suggested for implementation which are presented in the reports generated from the risk assessment process.

For traditional risk assessment, many tools are available [47, 48], however any tool developed for carrying out risk assessment should have the some of the following features:

- structured report generation to show the risk probabilities and their impacts,

- questionnaires and checklist to assess concerns related to compliance, policies, and best practices,

- list of threats and Security measures that can be used to suppress them, and

- software automation.

Qualitative vs. Quantitative Risk Assessment

Risk assessment involves the comprehensive identification of different threats an organization will face and evaluate their likelihood and impact. There are two primary methods of doing this: qualitative and quantitative. Qualitative risk assessment involves subjectively evaluating the identified risk's impact or likelihood using metrics like High, Medium, or Low. The categorization of the identified risks is usually done based on organizational policies and their understanding of decisions like what constitutes of a high impact risk on the organization. This process is simplistic in nature and allows organizations to develop a ball-park estimation of the overall threats faced by an organization.

Quantitative risk assessment in contrast involves numerical estimations and mathematical equations. This is done by using estimation techniques like Bayesian networks or Monte Carlo analysis to determine the probability of success for the likelihood or impact of different threats. Although, an inherent challenge in this technique is to able to accurately obtain the initial probability or risk values of the assets which are exposed to the identified threats. However, tools and techniques are available like Common Vulnerability Scoring System (CVSS) [54] which can be used as a foundation to customize the empirical estimation according to the needs of an organization. In doing so, quantitative risk assessment will allow for quantification based on different categories like (1) inherent difficulty of exploiting the threat, (2) priority given to different network security parameters, and (3) impact of initial security measures, and so on. Nonetheless, the goal of this method is to develop a framework that can lead to some actionable insights and not just provide an understanding of the threat level of the organization. The actionable insights may involve computing the security return on investment by performing cost-benefit tradeoff analysis, or evaluating the effectiveness of different security measures.

However, performing accurate and comprehensive quantitative risk assessment is ideally complex and costly. Therefore, it cannot be universally incorporated by small- or medium-scale organizations. A more pragmatic approach lies in using hybrid risk assessment methods that involves qualitative and quantitative techniques. Such a hybrid approach first will involve initially estimating and prioritizing identified threats using qualitative assessment measures. Next is to apply quantitative measures to design and develop actionable insights.

2.2   CRYPTOGRAPHIC OPERATIONS

2.2.1   HOMOMORPHIC ENCRYPTION

A function f is said to be homomorphic if, for an operation $\oplus$, it exhibits the following property:

$$f(a) \oplus f(b) = f(a \oplus b).$$

For example, the function $f(x) = x^2$ exhibits the homomorphic property for the multiplication ($*$) operation. The same function does not exhibit the homomorphic property for the addition ($+$) operation.

An encryption algorithm is called a Homomorphic Encryption algorithm or a Privacy Homomorphism, if it exhibits a similar behavior. Thus, for homomorphic encryption,

$$enc(a) \oplus enc(b) = enc(a \oplus b).$$

The operations that homomorphic encryption algorithms generally work with are the addition and the multiplication operations. If an algorithm supports one of these two operations it is called a partially homomorphic encryption algorithm. If it supports both operations it is called a fully homomorphic encryption algorithm. Encryption schemes, such as Paillier and Elgamal, are partially homomorphic encryption algorithms while Gentry's FHE [66] is an example of a fully homomorphic encryption algorithm. Homomorphic encryption offers us the ability to perform a restricted set of mathematical operations on encrypted data, thus eliminating the need to decrypt data. We will use homomorphic encryption in Chapter 5 when we present secure data aggregation algorithms that perform addition over the ciphertext on wireless sensors.

2.2.2   PAILLIER ENCRYPTION

The Paillier cryptosystem [106] is a public key encryption scheme based on the Decisional Composite Residuosity Assumption (DCRA). The DCRA states that given a composite number $n = n_1 \times n_2$, for primes n_1 and n_2, and an integer z, it is hard to find out whether there exists a number y such that $z \equiv y^n \bmod n$. A message $M \in \mathbb{Z}_n$ in Paillier encryption is encrypted as

$$C = g^M.r^n \bmod n^2,$$

where $g \in \mathbb{Z}_{n^2}^*$ is a public element, while $r \in \mathbb{Z}_n^*$ is chosen randomly by the encryptor. The ciphertext C can be decrypted as

$$M = L(C^\lambda \bmod n^2).\mu \bmod n,$$

where $\lambda = lcm(n_1 - 1, n_2 - 1)$, $\mu = (L(g^\lambda \bmod n^2))^{-1} \bmod n$ and $L(x) = (x - 1)/n$.

Paillier encryption exhibits the property of additive homomorphism which will be used in Chapter 6. Given C_{m_1} and C_{m_2}, paillier ciphertexts for m_1 and m_2, the ciphertext for $m_1 + m_2$ can be generated by multiplying C_{m_1} and C_{m_2}:

$$C_{m_1+m_2} = C_{m_1}.C_{m_2}.$$

$C_{m_1+m_2}$ can then be decrypted by the private key to retrieve $m_1 + m_2$.

2.2.3   ELLIPTIC CURVE CRYPTOGRAPHY

Public key cryptography, also known as asymmetric cryptography, is widely used for various purposes such as key distribution, secure communication, message signing, etc. However, traditional public key algorithms are generally energy-hungry and, therefore, not suitable for wireless sensors. Elliptic curve cryptography (ECC) is a form of public key cryptography that has emerged as an attractive and viable public key system for constrained environments.

ECC offers considerably greater security for a given key size, compared to the traditional public key systems. This means that smaller keys can be used for equivalent security which makes possible more compact implementations that are faster (Table 2.1). This also leads to less power consumption and in turn less heat production. Heat production is an often overlooked issue on wireless sensors that can affect the WSN operations in two different ways. First, heat is just

Table 2.1: Key size equivalence [67]

Symmetric Key Size (bits)	Public Key Size (bits)	ECC Key Size
80	1,024	160
112	2,048	224
128	3,072	256
192	7,680	384
256	15,360	521

precious energy on wireless sensors that is wasted; second, excessive heat can interfere with a sensor's immediate environment, causing faulty readings.

ECC is essentially a discrete log (DL) system where the security is based on the intractability of the elliptic curve discrete log problem in a finite field. An elliptic curve E_k over a field K can be defined as an equation of the form, $y^2 = x^3 + ax + b$ along with a point at infinity. Here, a and b are real numbers and $4a^3 + 27b^2 \neq 0 \pmod{p}$ where p is a prime number greater than 3.

For cryptographic purposes, a base point G is chosen and made public for use. A private key k is selected as a random integer and then the public key $P = k * G$ is calculated. The Elliptic Curve Discrete Log Problem (ECDLP) implies that it is hard to determine k given the public key P and the base point G. This problem is harder than the original discrete log problem and ECC systems are built around it.

In this book, ECC-based cryptography will be used in Chapter 5 for use in a secure data aggregation algorithm and again in Chapter 7 for use in proxy re-encryption.

2.2.4   KEY POLICY ATTRIBUTE-BASED ENCRYPTION

Key policy attribute-based encryption (KP-ABE) [105] is a concept in which each ciphertext is associated with a set of attributes. The access policy is defined by an access tree and the private key is generated based on this access tree, hence the name key policy. A cipher-text can only be decrypted with a key, if the attributes associated with the ciphertext satisfy the key's access tree. The KP-ABE [105] is composed of four algorithms, *Setup, Encryption, Key Generation, and Decryption*.

- *Setup:* The Setup algorithm defines the attributes in the system and outputs a public key *PK* and a master key *MK*. The public key *PK* is used for encryption while the master key *MK* along with *PK* is used to generate user keys.

- *Encryption:* The encryption algorithm takes as input a message m, a set of attributes γ, and the public key *PK*, and produces the cipher-text E.

- *Key Generation:* The key generation algorithm takes as input an access tree $\mathcal{T}$, the master key *MK*, and the public key *PK* to produce a secret key *SK*, such that *SK* can decrypt *E* iff $\mathcal{T}$ matches γ.

- *Decryption:* The decryption algorithm takes as input the ciphertext *E*, the decryption key *SK* and the public key *PK* and decrypts the ciphertext iff $\mathcal{T}$ matches γ, otherwise it produces $\perp$.

2.2.5   PROXY RE-ENCRYPTION

Proxy re-encryption allows third parties called proxies to re-encrypt a ciphertext generated using a secret key so that it can be decrypted using a different public key. Proxy re-encryption (PRE) was first presented in [120]. This scheme known as the BBS scheme is a very simple PRE scheme based on Elgamal encryption.

The concept of symmetric key proxy re-encryption (SRE) was explored by Syalim et al. in [119]. The symmetric proxy re-encryption scheme of [119] makes use of an All Or Nothing Transform (AONT) along with a symmetric cipher. AONT has the properties that it's output is pseudo-random, and the transformation of output back to input requires all the output blocks be in their correct positions. These properties are used along with a weak permutation cipher to develop a secure symmetric proxy re-encryption scheme. For algorithmic details, the reader can refer to [119]. Proxy re-encryption is used in Chapter 7 for ensuring the confidentiality of code transferred from the base station to sensor nodes.

2.3   OTHER MATHEMATICAL PRIMITIVES

2.3.1   BILINEAR MAPS

A bilinear map is a function that takes as input elements from two spaces and outputs an element of a third space. Let G_1 and G_T be cyclic groups of the prime order q. Typically, G_1 is an elliptic curve group and G_T is a finite field. We, therefore, denote G_1 using additive notation and G_T using multiplicative notation. Let P and Q be two generators of G_1. A bilinear map then is an injunctive function $e : G_1 \times G_1 \rightarrow G_T$, which has the following properties.

- Bilinearity: $\forall P, Q \in G_1, \forall a, b \in \mathbb{Z}_q^*$, such that we have $e(aP, bQ) = e(P, Q)^{ab}$.

- Non-degeneracy: If $P \neq 0$, then $e(P, P) \neq 1$.

- Computability: There exists an efficient algorithm for computing $e(P, Q), \forall P, Q \in G_1$.

Bilinear maps would be used in Chapter 6 for creating an access control scheme based on KP-ABE that works on aggregate data.

2.3.2 SHAMIR'S SECRET SHARING

In secret sharing, a data item δ is divided into n parts and distributed to n participants such that any k parts can be used to reconstruct the data. The knowledge of $k - 1$ parts does not provide any information useful for reconstructing δ.

To create shares of a data item δ using Shamir's [92] scheme, a prime number p is chosen such that $p > \max(\delta, n)$. A random $k - 1$ degree polynomial is then created, where the coefficients $a_1, \ldots, a_{k-1}$ are chosen randomly from a uniform distribution over the integers in $[0, p]$ and $a_0 = \delta$. This polynomial $\omega(x) = a_0 x_0 + a_1 x_1 + \ldots + a_{k-1} x_{k-1}$ is then sampled at points $x = 1, \ldots, n$ to create shares $\omega(1), \ldots, \omega(n)$. These shares are called secret shares.

Any k out of the n shares created this way can be used to create a $k - 1$ degree polynomial $\omega'(x)$ using a polynomial interpolation method such as Lagrange's. To reconstruct the original data $\omega'(x)$ is evaluated at $x = 0$.

Shamir's secret sharing is information theoretically secure. It is also additive homomorphic in nature, which means that shares from two secrets can be added to add the secret. This property of secret sharing is used in Chapter 5 to add data secured by the shares.

2.3.3 BLOOM FILTER

Bloom filter is a versatile space efficient probabilistic data structure used primarily to verify set membership [122]. The Bloom Filter consists of an array of bits, initialized to 0 and k different hash functions. To add an element to the Bloom Filter, the k hash functions are used to hash the element to obtain k array positions. These array positions are then set to 1. To verify an element's membership, it is hashed using the k hash functions and each resulting array position is checked. If any of the bits at these positions are found to be 0, the element is not a member. If all the bits are set to 1, it is a member. We have used Bloom Filter as a means of verification in Chapter 7.

CHAPTER 3

Sensor Cloud Architecture and Implementation

A sensor cloud is composed of virtual sensors built on top of physical wireless sensors. It can be dynamically provisioned or de-provisioned intuitively based on the demand of a user's application. This type of service provisioning approach brings about numerous advantages. First, it enables better sensor management capability by abstracting the physical infrastructure. The users with the help of virtual sensors can interact with, and control, their view of the relevant WSNs. The standard functions in sensor cloud involves different parameters such as region of interest, sampling frequency, latency, and security options for the transmitted data. Second, the data captured by WSNs can be shared among multiple users which reduces the overall cost of data collection from a system's point of view as well as from a user's point of view. The reusability of data in WSNs is transparent to the sensor cloud users, thus increasing efficiency by reducing the redundant data capture. Third, the sensor cloud system introduces physical infrastructure independence. Irrespective of the type of sensor nodes used in the available WSNs, users can deploy and execute their desired applications without having to worry about the low-level details of which type of motes and sensors are used and how to configure them; these details are automatically handled by the sensor cloud platform.

The definition of the sensor cloud system architecture presented here is different from what is presented in [29], which is an extension of the concept of Internet of Things (IoT) [28]. The IoT integrates the services provided by sensing devices with cloud computing over the Internet. The presented sensor cloud architecture here, on the contrary, is a cloud of virtual sensors built on top of physical sensors. It provisions and de-provisions virtual sensors to the users and aims to provide *sensing-as-a-service*. The authors in [28] discussed various components which constitute a sensor cloud system, its management, as well as the control flow of various components. Poolsappasit et al. [33] provided a layered architecture of a sensor cloud and outlined the security challenges of such a system. The abstraction of virtual sensors was first described in [30]. In [31], the authors designed the allocation algorithm to reduce the energy consumption while scheduling algorithms are designed to reduce the response time of sensors in multi-application scenario. Global sensor network (GSN) [32] is a middleware designed to rapidly deploy heterogeneous wireless sensors. GSN also relies on virtual sensors; however, unlike the concepts presented in [30] where virtual sensors are defined using classes, the virtual sensors in GSN are defined using XML. GSN, the closest approach to the presented sensor cloud system architecture here, offers a ready-to-use system which can integrate large number of WSNs. However, it cannot be classified as a sensor cloud as the purpose of a GSN is to provide efficient and flexible deployment for multiple standalone WSNs.

As a running example in the rest of this chapter, we will consider the following scenario. Traffic flow sensors are widely deployed in large numbers in places such as Washington DC and Ohio. These sensors are mounted on traffic lights and provide real-time traffic flow data. The sensed data can be utilized by drivers to improve the planning and management of their trips. If the traffic flow sensors are augmented with low-cost humidity and temperature sensors then they can be used to provide a customized and local view of temperature and heat index (calculated using temperature and humidity) data on-demand. The National Weather Service, on the other hand, uses a single weather station to collect environmental data for a large area, which may not represent the entire region accurately.

3.1   VIRTUAL SENSORS

To be able to understand the sensor cloud system architecture presented here, first and foremost, it is imperative to grasp the notion of *virtual sensor*. These virtual sensors are the foundational building blocks which bridges the gap between the users of the sensor cloud and its physical infrastructure (WSNs). Therefore, as mentioned earlier, virtual sensors lie at the crux of bringing about the numerous advantages introduced by the sensor cloud platform.

Virtual sensors are the sensors which are devised on top of actual physical sensors sensing various phenomena like temperature, humidity, air pressure, and so forth. They do not have an equipment footprint, that is they do not exist physically but provide a customized view to users utilizing notable concepts such as distribution and location transparency. In wireless sensors, the hardware is barely able to run multiple tasks simultaneously due to the inherent characteristics of their hardware which is resource constrained in terms of memory, energy, and processing power. Therefore, the question of being able to run multiple VMs like the machines in the traditional cloud computing is not a realistic consideration. To overcome this challenge, virtual sensors are implemented as an *image* in the software of the corresponding physical sensor(s). The virtual sensors contain metadata about the physical sensor(s) and the user currently holding that virtual sensor. Additionally, the virtual sensor may have some data processing code, which can be used to process data in response to complex queries from the user. The virtual sensors are implemented in four different configurations which are listed as follows.

One-to-Many.   One physical sensor corresponds to many virtual sensors. While the virtual image is owned by individual users, the underlying physical sensor is shared among all the virtual sensors accessing it. The sampling duration (time period of service provisioning) and frequency (rate at sensing operations are carried out the sensor motes) of the physical sensor are computed by taking into account all the users and is re-evaluated when a new user joins or an existing user leaves the system, thereby making the service provisioning and de-provisioning process both dynamic and intuitive. This configuration can be seen in Figure 3.1.

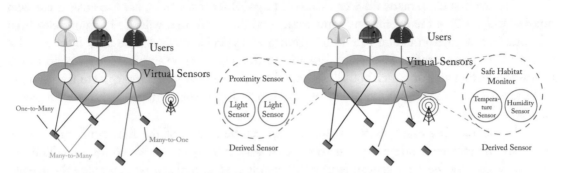

Figure 3.1: Virtual sensor configurations.

Many-to-One.   The geographical area is divided into regions and each region may have one or multiple physical sensors and sensor networks. When a user service requests aggregated data of a specific sensing phenomena from a region, all underlying WSNs are switched to active mode with the respective phenomena enabled and the user is provided with aggregated data from these WSNs. The sampling time interval at which all underlying sensors sense is equal to the sampling time interval requested by the user. This configuration can be used to provide fault tolerance if the underlying physical sensors fail. A virtual sensor communicates with a number of underlying physical sensors and it shows the aggregate view of the data to the user. When physical sensors fail, this failure is captured by the WSN-facing layer of the sensor cloud and communicated with the virtual sensor. The required data may be gathered from a working sensor which provides data within the quality of service (QoS) limits. Adapting to a change in topology thus is handled by the virtual sensor and the WSN-facing layer and is transparent to the user.

Many-to-Many.   This configuration is a combination of the One-to-Many and the Many-to-One configurations. A physical sensor may correspond to many virtual sensors while it may also be a part of a network which provides aggregate data for a single virtual sensor, as shown in Figure 3.1.

Derived.   This is a versatile configuration of virtual sensors which is derived from a combination of multiple physical sensors. This configuration can be seen as a generalization of the above three configurations. However, the difference lies in the types of physical sensors a virtual sensor communicates with. While in the derived configuration the virtual sensor communicates with multiple sensor types, in the other three configurations the virtual sensor communicates with the same type of physical sensors. In this regard, the derived sensors can be utilized in two ways. First, to virtually sense complex phenomenon and, second, to enable the substitution of the physical sensors that are required by the user's application but may not be available in the physical infrastructure. These two scenarios can be elaborated with the help of the following examples.

For the first case, many different kinds of physical sensors can be used to answer complex queries such as "Are the overall environmental conditions safe in a wildlife habitat"? Readings of a number of environmental conditions from the physical sensors can be used by the virtual sensor to compute a safety level value and answer the query. An example of the second case is to use the virtual sensor which could use data from light sensors and interpolate the readings and the variance in the light intensity to act as a proximity sensor. Figure 3.1 shows examples of derived sensors.

In our running example, the virtual sensor can be in any of the four configurations. A user may interact with one particular traffic flow sensor to assess traffic conditions. Multiple users may also use the same sensor. A user may configure a virtual sensor to provide the average temperature of a region which may involve multiple sensors. A user may also configure a derived virtual sensor to calculate heat index from temperature and humidity data.

3.2   SENSOR CLOUD ARCHITECTURE

The sensor cloud system architecture has three prominent layers—client-centric layer, middleware layer, and sensor-centric layer. The client-centric layer connects the users to the sensor cloud, while the middleware layer performs service negotiation, provisioning, and maintenance of virtual sensors, and communication of data from the sensor-centric layer to the client-centric layer. The sensor-centric layer deals with the physical wireless sensors and their maintenance as well as routing of data and commands. The high-level conceptual system model of the sensor cloud is shown on the left side of Figure 3.2. From an implementation point of view, the layered architecture can be condensed to the block diagram shown on the right side of Figure 3.2. Each block comprises of one or more related functionalities in the layered architecture. The block diagram representation is a more implementation friendly illustration of these functionalities. In the remainder of this section, we describe the layered sensor cloud system architecture as it is represented in Figure 3.2.

3.2.1   CLIENT-CENTRIC LAYER

The client-centric layer acts as the gateway between the sensor cloud and the user. It is a collection of components which facilitates and manages the interactions between the user and the core of the sensor cloud, i.e., the virtual sensors. The client-centric layer is composed of the user interface, session management, membership management, and the user repository components.

User interface is the graphical front end that enables users to communicate with sensor cloud. The interface, facilitated as a web application, allows users to specify parameters such as region(s) of interest, sensing phenomena, sampling frequency, sensing duration, and data delivery security mode (encrypted or unencrypted). The request along with the parameters is parsed by the web application and communicated to the back-end application server. Session

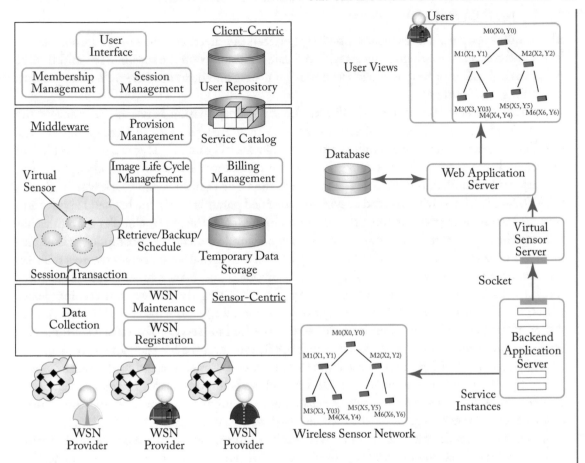

Figure 3.2: Layered sensor cloud architecture and block diagram representation.

management component handles the secure creation, management and termination of sessions between the middleware and the user. The membership management component of the client-centric layer takes care of the authorization of users and provision of access to their authorized services. Finally, the user repository component of the client-centric layer stores the detailed information of users in the system. It stores account credentials, personal information, payment history, billing information, and so forth, of the users. Along with that, it also keeps track of data sent by WSNs and accessed by the end users.

In our example, the client-centric layer will expose a web application-based GUI showing the locations of the available sensors to the user. The user can create virtual sensors based on selected regions. The information about the selected region, sampling frequency and duration, and QoS agreement would be sent to the middleware layer.

3.2.2   MIDDLEWARE LAYER

The middleware layer acts as the intermediary between the client-centric and the sensor-centric layers and connects the client requests with the data collected from the sensors. The middleware performs a number of functions such as provision management, image life-cycle management, and billing management.

The provision management facilitates the service negotiation between the user and the sensor cloud, and provides virtual sensors for each incoming request. This component of the middleware resides in the web application server block of Figure 3.2. It receives requests from UIs of multiple users with their functional service parameters and security mode option. If a request can be fulfilled, the control is passed on to the virtual sensor server block which triggers the WSNs associated with selected region and specified parameters via the backend application server. Once an agreement between the user's requirement and the sensor cloud's capabilities has been reached, the virtual sensors are created. Image lifecycle management component is implemented in the virtual sensor server block of Figure 3.2. The virtual sensor server receives requests for instances from the provision management component and takes care of creating instances for the virtual sensors provisioned for the users. After creating the instances, it communicates with the backend application server to request a corresponding service instance. Each service instance of the backend application is implemented as RESTful service. The virtual sensor server mediates the communication between the web application and the backend application and ultimately displays a different set of sensor data to each end user connected to the sensor cloud. The billing management component of the sensor cloud is responsible for keeping track of each user session with respect to the types of sensors used, number of sensor used and the security mode of sensor usage to generate the invoice based on a service agreement beforehand.

In our example, the middleware layer receives information about the user session from the user centric layer and creates the virtual sensor. The virtual sensor configuration is decided based on the type of data the user wants, users region of interest and the agreed upon QoS. In case multiple users request information from the same sensors (for example, traffic information from the same location), the requests are consolidated by the middleware.

3.2.3   SENSOR-CENTRIC LAYER

The sensor-centric layer directly communicates with the physical sensors using the WSN registration, WSN maintenance and the data collection components. When a network owner wishes to provide service through the sensor cloud interface, they need to register their WSNs. The physical sensors and their capabilities are verified by the sensor cloud and the physical sensors are expected to provide their location information. For the current sensor cloud system architecture, mobile sensor motes are not taken into consideration, therefore, pre-deployed location information in the form of longitude, latitude, region-id, and cluster-id should be sufficient to store location information. The collected information in the WSN registration phase is used in cataloging the information about physical and later virtual sensors. Once a WSN is registered, the network owner is in charge of keeping the physical sensors in good health. The registration binds the WSN owner and the sensor cloud in a bi-directional trust relationship, where the WSN owner is expected to provide accurate non-tampered sensor readings, while the sensor cloud is expected to correctly provide compensation for the received sensor readings. The trust between the two parties can be enhanced by using secure and trusted data collection and aggregation techniques on the WSN side. Further, data anonymization can be implemented, and computations can be carried on encrypted data which can be performed on the cloud side.

The WSN maintenance component provides interoperability of the heterogeneous mote platforms, periodically monitoring the health of each mote in the sensor cloud. It also provides synchronization between sensors, and collects metadata information about the motes and the networks. To manage non-interoperability, we install a GumStix at the junction of two or more incompatible WSNs. The GumStix is hooked up with a number of different motes at different ports and it collects data from one port and transmits it to other ports as required by the application. For monitoring the health of the motes and collecting metadata information, the WSN maintenance component periodically pings each network. The motes in the network reply with data packets consisting of information such as their battery level, last active connection, location, region, and so on. While the challenge of synchronization between the networks and sensors is not actively addressed, it can be done so by allowing the global time information to be piggy backed on the ping packets responsible for WSN maintenance. This would provide time synchronization between the networks and the sensors. To provide a finer-grained synchronization we can also use the powerful GumStix nodes.

The data collection component of the sensor-centric layer connects the system directly to the WSNs. Each WSN is connected to the data collection component through a base station. The data collection component which resides in the backend application server runs one service instance for each base station. The service instance opens two dedicated ports—one to communicate with the base station and one to communicate with the associated virtual sensor on the virtual sensor server. The sensor-centric layer in our example receives query packet from the middleware and replies with the requested data. The routing protocol, fault tolerance, and other network-related issues are handled by this layer.

3.3   SOFTWARE DESIGN

The sensor cloud is a multi-tiered client server software architecture with each layer logically separated from the other. The sensor-centric layer is the data tier. This layer consists of physical wireless sensors that generate real-time data. The middleware is the application or the logic tier which controls the data collection, while the presentation tier is represented by the client-centric layer.

From a software developer's point of view, there are two different facets to developing a sensor cloud. The system side of the sensor cloud and the sensor side of the sensor cloud. The system side consists of the client-centric layer and the middleware (the presentation and application tiers) and is basically used to manage the physical resources. The view of the end user may vary depending on the application. The heart of any sensor cloud application, though, is going to be the middleware layer. The middleware is expected to be flexible enough to handle the issues when physical and virtual sensors are scaled up and down. The middleware layer is responsible for aggregating the user requirements and re-direct data according to these requirements from the sensors to the users, in addition to handling the tasks discussed earlier.

The second facet of a sensor cloud is the physical sensors side which consists of the sensor-centric layer. WSNs are distributed networks, where a number of physically separate entities work together toward a common goal (in this case, generating data according to the user's requirements). WSNs face the same general issues faced by distributed systems such as synchronization, fault tolerance, and security. While developing for WSNs, developers need to keep these issues in mind, in addition to the wireless sensor specific constraints such as low bandwidth, low processing power, and a finite energy source.

3.4   QOS IN SENSOR CLOUD

Quality of service is managed at two levels in a sensor cloud, first at the sensor-centric layer and second at the virtual sensors level. Network-related issues, such as responding to node failures, network partitioning, and packet losses, are handled by the sensor-centric layer. The virtual sensor layer then works on top of the services provided by the sensor-centric layer to manage QoS parameters such as reliability, data accuracy, and coverage on top of the network layer. As explained earlier, the many-to-one virtual sensor configuration can be used to provide reliability of data by making use of data aggregation. The accuracy of data in such cases depends upon the spatial and temporal correlation between the nearby sensors. The virtual sensor layer makes use of the correlations to retrieve data within the QoS limits from nearby sensors. The virtual sensor layer can also switch between configurations such as from one-to-many to many-to-many to provide data to multiple users within the specified QoS limits.

3.5   IMPLEMENTATION

In this section, we describe the translation of the conceptual sensor cloud architecture, as shown in Figure 3.2 to its corresponding physical model for implementation purposes.

3.5.1   SYSTEM SETUP

The sensor cloud web application was implemented using the *MEAN* framework. Linux platforms (Ubuntu 14.04 LTS) were used to host backend base station servers in the sensor-centric layer and the middleware. The front end was built using HTML5, CSS3, and Angular. The middleware was built using server side NodeJS, which is supported by a Mongo—DB database. The backend base station servers were built using RESTful web services written in Java JAX-RS. Communication between the middleware and backend base stations was done via web service calls. The wireless sensors used for the implementation prototype were TelosB motes, each equipped with humidity, temperature, light intensity, Infra Red sensor, and vRef sensor [21]. The sensors were programmed using TinyOS 2.2.x. Figure 3.3 shows the overview for the system setup of sensor cloud web application.

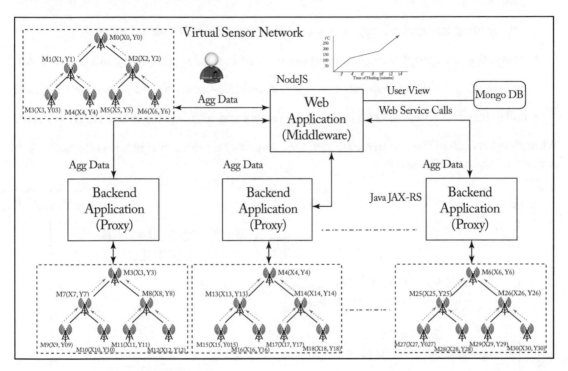

Figure 3.3: Sensor cloud system setup.

3.5.2   MIDDLEWARE IMPLEMENTATION DETAILS

The Sensor Cloud middleware system is implemented using a NodeJS platform. The inherent advantage of using NodeJS lies in its faster execution and response time. The middleware is solely responsible for managing both user sessions (sending requested data to each user) and participating WSNs (fetching data from various WSNs) and these operations are highly time consuming. As such, the NodeJS middleware will scale well with the increase in the number of WSNs, and the number of users making simultaneous service requests in the sensor cloud. Another advantage of NodeJS can be attributed to its mode of operation using JSON data format. This makes interactions with web services and database calls faster and easier to maintain. As the middleware has to frequently interact with the sensor-centric layer (heterogeneous in nature) using web services, NodeJS makes the operation much faster and efficient as compared to JAVA.

Currently, the implemented sensor cloud middleware supports the following basic functionalities:

- registering a backend base station server; hosting single or multiple WSNs, with the sensor cloud;

- determining if a selected region by a user has physical WSN sensing support;

- computing and propagating sampling duration and sampling frequency values to a WSN;

- transmitting activation requests to single- or multiple-backend base station server(s); and

- collecting data from backend base station servers and transmitting it to authorized users.

Work flow for each of these aforementioned functionalities are described below and summarized in the flow chart in Figure 3.4.

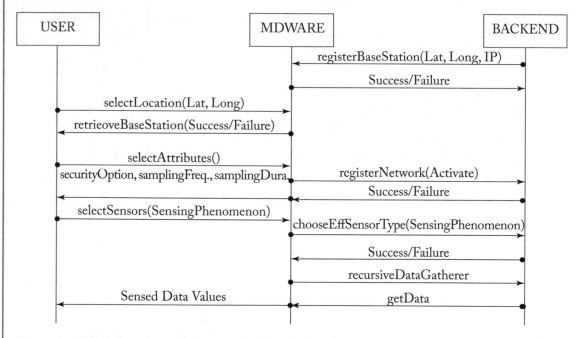

Figure 3.4: Workflow for middleware implementation.

1. *Registration*: This method accepts a request from a backend base station hosting a WSN(s). The request contains the geographical latitude and longitude values of the deployed network and IP address of the machine where the backend base station server is hosted. When middleware receives such a request it stores this information in the database and a JSON response is sent with success field indicating *TRUE*, if the operation is successful and

FALSE otherwise. *registerBaseStation* of *DatabaseConnectivity* Java Script is called for storing the WSN information provided by the backend base station server.

2. *Location*: This method accepts the user request from the frontend which contains the requested location information. For a selected location, this method returns a JSON response to the front end which contains a *TRUE* for the success field if there is a WSN service location present in the database. Otherwise, it returns a *FALSE* in the success field. This method makes a function call to *retriveBaseStation*, present in *DatabaseConnectivity* Java Script file, to retrieve WSN information from the database. If WSN information is found, it is stored in the session and a *TRUE* is returned but if WSN information is not found then a *FALSE* is returned.

3. *RetriveBaseStation*: This method accepts latitude and longitude values of the location selected by a user. For the selected location, database is queried to find the WSN information. If a WSN is found, the information is returned otherwise a null object is returned.

4. *SelectAttributes*: This method accepts request from the frontend that contains the values of security option (secured or unsecured), sampling duration, and sampling frequency from a user. It also accesses the WSN information and updates the session information with the security option, sampling frequency, and sampling duration. It then calls *registerNetwork* to send activation command to the selected WSN(s). Based on the success of the activation, a *TRUE* or *FALSE* is set in the success field and sent as JSON object to the frontend as response.

5. *RegisterNetwork*: This method accepts the sampling duration, sampling frequency, security type, and WSN information. On receiving the information, it calls a web service to activate the WSN. After activating a WSN, sampling duration and sampling frequency are updated in the database for the location, if already present. The updated sampling duration is the maximum of the current sampling duration and that of the new request. Sampling frequency is the minimum of the current and that of the new request. If a request for a WSN is not present in the database, sampling duration and sampling frequency of the new request is stored in the database.

6. *SelectSensors*: This method accepts the user request from the frontend which contains security and effective sensor type information. For each request, WSN information stored in the session is also retrieved from the user request. A JSON object is returned with success field as *TRUE*, if the activation is successful and *FALSE*, if the activation is unsuccessful. This method calls *chooseEffectiveSensorType* method to activate necessary WSNs.

7. *ChooseEffectiveSensorType*: This method accepts WSN information, security, and effective sensor-type information. For each WSN, a web service call is made to change the sensing capabilities by transmitting the security and effective sensor-type information. When the

sensing capabilities information is successfully sent, a function call is made to *recursive-DataGatherFunction* to start fetching data.

8. *RecursiveDataGatherFunction*: This function accepts latitude, longitude and other required WSN operational information. A database call is made to check if the sampling duration time has expired for a given WSN. If the sampling duration has expired, the recursive call is terminated and a web service call is made to the WSN to deactivate the sensors nodes. If sampling duration time has not expired, a web service call is made to fetch the data from the WSN. After fetching the data, it is stored in the database and a recursive call is made after a timeout duration which is specified by sampling frequency time interval. This recursive function call is independent from other methods in the middleware and mimics the functionality of a thread in the JAVA programming. It fetches data from the WSN and stores it to the hadoop/mongo database irrespective if the user is currently active. This feature is useful in allowing the users to query past data.

9. *GetData*: This method accepts a user request from the frontend to retrieve data from the requested WSN(s). This method will return the current values of requested sensor information in JSON format to the frontend. First, this method sets a timeout for the specified sampling frequency time interval in the user request. This timeout will make the user wait for the specified sampling frequency time duration before it returns a value such that continuous request from the user do not congest the network. After the timeout duration, this method calls the *DataValues* method to fetch current real time values which are compiled into a JSON object and returned to the user.

10. *DataValues*: This method accepts WSN information and effective sensor-type information as requested by the user. Based on the WSN information, database is queried to retrieve latest values of the sensors for the requested WSN's. Result of the database query is returned.

3.5.3   BACKEND BASE STATION SERVER IMPLEMENTATION DETAILS

Backend base station servers are implemented in JAVA as there is no interoperability between tinyOS [23] and NodeJS [24]. Backend base station servers perform four basic functionalities.

- Register the base station of a physical WSN.

- Activate the sensors motes. Deactivating the sensors motes.

- Transmit the change in sensing capabilities.

- Fetch data from physical WSNs.

Work flows for each of these functionalities are described below.

1. *Register*: This method is called during the startup of the WSN server. It does not take any parameters as input but, it calls the web service of the middleware by passing latitude, longitude, and IP address information to register this base station. We hard-coded the latitude and longitude values for now as the sensors do not have location sensing capabilities. IP address is dynamically accessed from the wired machine which hosts the server.

2. *AddFrequency*: This method accepts sampling frequency as input and responds back with a *TRUE* or *FALSE* if sensor activation is successful or unsuccessful respectively. First, this method detects all sensors that are connected to the WSN. Then, it sets up the routing, TDMA and aggregation protocols for the sensors that are present in the WSN. If all the setup procedure is successful, it returns a *TRUE* value. It later sends out the sampling frequency to all sensors nodes.

3. *RemoveFrequency*: This method, when invoked, sends out command to all the sensors to stop sensing and go to sleep mode.

4. *SetEffectiveSensorType*: This method accepts the new sensing parameters that are to be sensed by the sensors nodes. It then sends out this received sensing capabilities information to all the sensors in the WSN. If the sensors are previously sensing, they would reset and start sensing again. On the other hand, if the sensors are not sensing, they will start sensing after they have received the sensing capability information. Once the sensors start sensing, they are programmed to send data to the serial forwarder object which later forwards it to WSN object.

5. *GetData*: This method accepts a request from the middleware and responds to this request with the new sensed values. These data values are obtained from the WSN object which keeps updating itself whenever the sensor nodes transmit a new set of sensed data value.

3.5.4   DATA STREAMING FOR MULTI-USER ENVIRONMENT

The virtual sensor model can be used to effectively support a multi-user environment. A single wireless sensor is used to provide data for multiple users, where the users may demand data at varying frequency and of different phenomena. When the web application server in Figure 3.2 receives user requests, they are transferred to the virtual sensor server. The virtual sensor server performs the mapping between the user's virtual sensors and the physical sensors. If multiple virtual sensors correspond to a physical sensor, the virtual sensor server combines the request by combining the sampling duration, sampling frequency and the sensing phenomena. The combined request is then forwarded to the appropriate service instance of the backend server. The service instance communicates with the WSN and collects data. Data from each WSN is sampled at the minimum frequency of all requests. This data is time-stamped with the local time and stored in the database and displayed to the user at the requested frequency. Also, users can select data from multiple base stations at a time, either by selecting location which includes multiple

WSNs or by selecting multiple such locations. The data will then be aggregated periodically at the requested frequency on the basis of hierarchy of the selected locations.

The sensor cloud system reduces the cost of receiving data from WSNs to a great extent. Since wireless sensor motes are embedded with various types of sensors, a multi-user environment provides flexibility to multiple users to receive data of different phenomena at different sampling time intervals from same WSN through data sharing. For each WSN, multiple users can select one or more sensing phenomena from the available ones. A superset of all phenomena requested by multiple users is taken into consideration and only those phenomena are enabled, while all others are disabled.

This further saves the energy of the sensors by minimizing volume of data sent from motes to application for unwanted phenomena. Sensing phenomenon and sampling time interval selection are dynamic in nature such that when a user joins the network or an existing user wishes to discontinue service, then phenomenon to be enabled or disabled and sampling time interval for the selected WSNs are reevaluated and reconfigured. Additionally, when the services of a WSN are not requested by any users, all motes belonging to that network are pushed to idle mode.

An architecture to enable multi-user access to single WSN is shown in Figure 3.5. End user requests middleware for data from a specific WSN with required phenomenon and sampling time interval. Middleware sends these requests to the backend base station server of the associated WSN where the base station server validates the incoming request. If a WSN is in idle mode, it is moved to an active mode and network undergoes topology discovery phase. In this phase, the base station sends out HELLO messages trying to discover its child node and as such, a WSN is formed in accordance to a hierarchical tree structure. Once the topology discovery phase finishes successfully, the sensor nodes senses the data for the requested phenomena and sends it to the base station sensor node (root node) which forwards it to the base station server. The base station server then forwards the data to the middleware. Depending on a user's request, middleware displays data to each user according to their requested frequency, time duration, and sensing phenomenon.

3.5.5   VIRTUAL SENSOR IMPLEMENTATION

In this section, we describe the usage of many-to-one virtual sensors; one-to-many, many-to-many, and derived sensors are also implemented in a similar fashion. The geographical area covered by the sensor cloud is divided into regions which are arranged hierarchically. Data from various networks in a region is aggregated. The aggregated data will be based on the user's selection of the region and hierarchy of selected region. The scheme can be understood well by visualizing a network of WSNs shown in Figure 3.6. The topology in Figure 3.6 shows the hierarchy of WSNs, where each intermediate node and root node is a network in itself and can have any number of children. In this example, we assume that an end user needs data from region 1 which is a virtual node. Similarly, all intermediate nodes (2, 3), and root node in the hierarchy

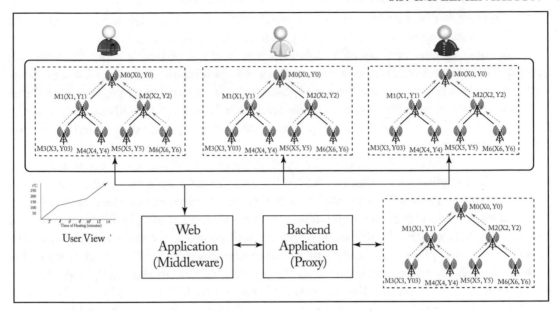

Figure 3.5: Multi-user access to a WSN.

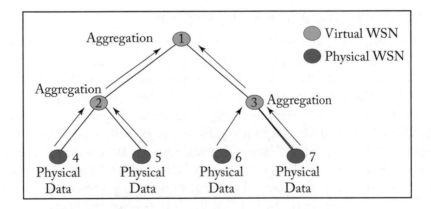

Figure 3.6: Architecture of virtual sensors.

are virtual nodes. On the contrary, all leaf level nodes (4, 5, 6, 7) represent physical WSNs. Thus, we can call this topology a network of VSNs and WSNs. The hierarchy information stored in data table is static and is required while aggregating the information at intermediate and root level nodes. On the other hand, a hierarchy data object is created for each user request. Once a many-to-one mapping of the above virtual sensor is obtained, the virtual sensor server sends a request to the backend application server, which then forwards the request to the concerned

WSNs. Once all WSNs providing data to selected VSNs in the hierarchy are switched on the backend application server starts relaying data which is parsed according to the VSN hierarchy.

We consider tree topology of WSNs to represent cloud of sensor networks, internally aggregated data from WSN would be represented as the data for a specific virtual sensor. These virtual sensors will be representing the geographical region(s) selected by a user. Figure 3.6 shows the hierarchy of the network where each leaf level node represents a WSN and other nodes represent regions. The order of data aggregation for a virtual sensor node depends on its position in the tree topology.

Architecture of virtual sensors is displayed in Figure 3.6. When user needs data from a region which is a collection of multiple WSNs, the middleware acknowledges the requests and forwards activation message to associated base station servers. Base stations behave in the same manner as explained in multi-user environment, however, because multiple WSNs are involved in this case, data from these WSNs are aggregated and displayed to the user as per their request, i.e., selected sensing phenomena and sampling time frequency and duration. Middleware takes up the responsibility to aggregate data by making a data request to the multiple WSNs that are involved in such scenarios. The advantage of this approach is evident by the fact that it provides location transparency to users by hiding minute details about the physical locations of the sensors, while dispensing their service requests.

3.5.6   TIME MODEL FOR VIRTUAL SENSORS

Consider two WSNs, WSN_1 and WSN_2, sensing and sending data to the middleware through backend applications. $User_1$ needs aggregated data from WSN_1 and WSN_2 at a time interval of 15 min; and $User_2$ needs data from WSN_2 at 10 min time interval. In this case, physically WSN_1 will sense data at frequency 15 min because WSN_1 has to serve only one user at a time interval of 15 min. On the other hand, for WSN_2 there are two requests from $User_1$ and $User_2$ at time intervals of 15 and 10 min, respectively. Hence, WSN_2 will sense data at the minimum frequency among all the requests (minimum interval between 10 min and 15 min is 10 min). While displaying data to the User1, latest data from WSN_1 and WSN_2, at the interval of 15 min will be aggregated and shown. So $User_1$ will be displayed the aggregated data from WSN_1 and WSN_2 every 15 min, irrespective of the fact that WSN_2 is sensing data at the 10-min interval. Although WSNs are physically sensing data at different intervals, the end user receives the data at the interval that they have selected. We refer to this scenario as virtualization. The end user is transparent to the frequencies at which WSNs are running. With the compromise of some accuracy, the sensor cloud is capable of provisioning data for more than one user at the same instance of time. Figure 3.7 shows the Time model for the sensor cloud [27].

3.6   SUMMARY

Sensor clouds aim to take the burden of deploying and managing the network away from users by acting as a mediator between the users and the WSNs and providing sensing as a service. In

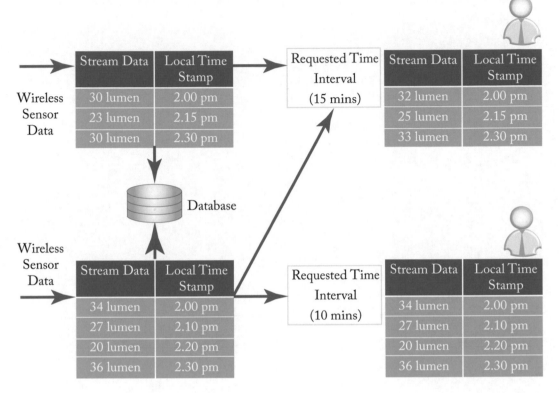

Figure 3.7: Time model for sensor cloud.

this chapter, we described the layered architecture of the sensor cloud system both on a conceptual and physical level which allows sensor networks spread in wide geographical areas to be connected and used as a single entity. It further allows the usage of sensor networks by multiple users at the same time by virtualizing the resources in software. Virtualization also helps in creating what we term as derived sensors from heterogeneous data streams.

CHAPTER 4

Risk Assessment in a Sensor Cloud

4.1 INTRODUCTION

A sensor cloud [19], as discussed in the previous chapter, consists of several different WSNs provided as a service to users through the sensor cloud middleware platform. The WSNs are comprised of low-cost nodes deployed in an ad-hoc fashion over a large area [33, 34] to collect attributes such as temperature, humidity, and other sensitive environmental data, as requested by a user's application. These deployments maybe done in hostile environments wherein they are not physically monitored for long time intervals. Additionally, with the integration of WSNs having different ownership entities under a sensor cloud platform, running a variety of user applications, increases the likelihood of attacks. This is because given the resource constrained nature of sensor motes (in terms of memory, processing power, and energy), the well-known traditional security mechanisms like strong encryption and privacy preserving measures cannot be adopted and implemented.

As such, there is a need to design and develop optimal security measures. However, before we can do that, we need to understand what kind of different security threats a sensor cloud platform composed of heterogeneous WSNs are prone to, and what will be the impact of the incorporated security measures in terms of its resources utilization vs. the extent of security it will provide. To satisfactorily answer such questions, performing security risk assessments for a sensor cloud platform is a pre-requisite. Risk assessment mechanisms will help to estimate the likelihood and impact of different known attacks on these WSNs in a sensor cloud.

The output generated by a risk assessment mechanism will help us to answer questions such as following.

- How can we better secure these WSNs in a sensor cloud?

- What are the possible attacks on our network and their risk level in the presence or absence of various security measures?

This will help in strengthening the network security before we integrate a WSN in a sensor cloud.

Risk assessment mechanisms have well-founded concepts in the literature and their applications can be found in several different domains—software projects, traditional wired networks, and even for the personnels who will be interacting with our systems. Most notably, the risk assessment mechanisms are well documented by governing bodies such as National Institute of Standards and Technology (NIST) in their special publications SP 800-30 for conducting risk assessment for information technology systems, and SP 800-37 for applying risk management frameworks to federal information systems. Although these works pioneered the standardization of early risk assessment mechanisms, they were further improved to incorporate a key component—understanding the cause-consequence relationship between the different security threats that are present and the feasible attacks that can exploit these security threats. A risk assessment mechanism adopted for a sensor cloud platform needs to be equipped with this cause-consequence identification characteristic.

To better understand the implications of this requirement, let us consider the following scenario. We want to assess the impact and likelihood for different security threats and attacks that can exploit them in the WSNs of a sensor cloud. To do so, we adopt a traditional risk assessment mechanism to understand the security risks confronted by the WSNs. Based on the initially chosen security measures and the corresponding risk assessment analysis, let us assume we conclude that attacks such as Sinkhole [36] is not feasible on the WSNs. However, such an assessment is carried out in a stand-alone fashion, i.e., by considering the WSNs in isolation. This is not the case in a sensor cloud domain since given its architecture, the WSNs will interact with different clients serving their application codes and requirements through the middleware and client-centric layers. As such, a malicious client's application may inject harmful executable codes via the middleware into the physical sensor motes which can lead to a node subversion attack [35]. Thereafter, an adversary can use the compromised node to break the authentication scheme. This will prompt the execution of other degenerate attacks like Sinkhole [36] or Sybil [37].

To incorporate cause-consequence analysis techniques in a risk assessment mechanism, one of the well-known methodologies is to perform risk assessment using the attack graph structure. Attack graphs are graphical structures with a bottom-up assessment approach in which the root node is an asset (for example, access privilege to an admin system) that we are trying to assess based on the identified vulnerabilities in various systems it may interact with. These systems are depicted as the intermediate and the leaf nodes of the attack graph structure. Attack graphs are well established for wired networks [40, 41] and [42], and we discuss them further in Section 4.2.1. As these ideas are researched and adapted for WSNs within a sensor cloud framework, they help us understand the cause-consequence relationship between attacks and identify ways in which a particular WSN security parameter may be exploited.

However, attack graphs implemented in the traditional domains cannot be applied as-is in the domain of WSNs in a sensor cloud because the characteristics of infrastructure and their functionality is inherently different from each other. When we generate an attack graph for a wired network, the network is scanned using a vulnerability scanner tool such as Nessus [47]. These scanners detect the list of vulnerabilities as shown in Figure 4.1, present on each system

High (9.3)	63228	MS 12-081: Vulnerability in Windows file handling component could allow remote code execution (2758857)
High (9.3)	63229	MS 12-082: Vulnerability in DirectPlay could allow remote code execution (2770660)
High (7.2)	35453	Microsoft Windows update reboot required
High (7.2)	63155	Microsoft Windows unquoted service path enumeration
Medium (6.4)	51192	SSL Certificate cannot be trusted

Figure 4.1: List of vulnerabilities—Nessus Network Scan.

in a wired network. This information is then parsed into an attack graph-generating tool [49]. However, no comparable vulnerability scanning tool currently exists for WSNs.

Hypothetically, if we assume that such vulnerability scanning tool existed for WSNs, generating such a vulnerability list (Figure 4.1) will not be sufficient for the sake of generating an attack graph. This is because sensor nodes in a WSN collaborate to achieve a common goal and suffer from inherent resource limitations. These limitations are the primary cause of WSN vulnerabilities as they do not permit the application of desired security protocols to safeguard the network. Hence the vulnerability list will be identical for all sensor nodes belonging to a WSN and no concrete conclusions could be formulated. Thus, rather than focusing on vulnerabilities in a sensor node or a sensor network, we can focus on the feasibility of attacks on a particular WSN in a sensor cloud. The successful execution of different attacks will vary according to security measures used, tasks being carried out, and deployed environment of a WSN. This is the principle idea of the attack graph structure we propose for our risk assessment framework to be used for WSNs in a sensor cloud. Nonetheless, establishing cause-consequence relationships alone is not sufficient. In order for a security administrator to develop actionable insights, for example, judging the efficiency and effectiveness of chosen security measures by performing a cost-benefit tradeoff analysis, a quantitative perspective of the risk assessment output is required. Instead of saying that a sensor network is highly secure because impact and likelihood of a Sinkhole attack is low, we should be able to enumerate the extent of this security by numerically estimating the likelihood and impact of an attack or a set of attacks. In this regard, we merge our proposed attack graph model with the concepts of Bayesian networks [45] and adopt the severity rating system for vulnerabilities as proposed by The National Institute of Standards and Technology's (NIST) Common Vulnerability Scoring System (CVSS) [44]. Hence, we can numerically evaluate the risk to various security parameters like confidentiality, integrity, or availability within a WSN.

Additionally, given the hostile and unpredictable environments in which WSNs are deployed, the complete safety of a WSN in a sensor cloud is an idealistic scenario. However, being able to predict the degradation of WSN security parameters such as confidentiality, integrity, and availability [38], and taking appropriate precautions such as redeploying the WSN using improved security measures or carrying out maintenance and repair activities is always a better alternative. Therefore, in addition to quantitative perspectives, time frames with respect to a WSN uptime is another useful estimate that help us reason better as a security administrator. For this purpose, we adapt risk level estimations modeled as a continuous-time Markov process in Houmb's Misuse frequency model [46] for WSNs. Houmb et al. proposed the methodologies of risk level estimations using the exploitation frequency and impact of vulnerabilities in a wired network. We adopted these concepts to identify the metrics necessary to compute net threat level to the root node of our proposed attack graph model when it is represented as a Bayesian network. This gives a degree of diversification and uniqueness to the WSNs with respect to quantitatively analyzing our attack graphs. This will help in predicting the degradation of security parameters within a WSN and schedule for time frames to carry out either maintenance or repair activities in the largely unattended WSNs.

Summarizing the above factors, carrying out cause-consequence-based quantitative risk assessment in WSNs under a sensor cloud is challenging and, hence, we address these challenges with the help of the following topics in this chapter.

- We formulate attack graphs to depict the logical correlation between the attacks on WSNs [50] (Section 4.2.1). These attack graphs will be then used to analyze how the attacks can exploit the WSN security parameters [38].

- We depict the attack graphs as a Bayesian network (Section 4.2.2), quantifying the likelihood and impact of the attacks in a WSN and the net threat level to WSN security parameters.

- We compute time frames predicting the degradation of WSN security parameters by modeling our risk level estimations as a continuous-time Markov process (Section 4.2.3). Given these time frames we can take precautionary measures and perform maintenance in an unattended WSNs before they reach an irreparable state.

4.2   RISK ASSESSMENT FRAMEWORK FOR WSN IN A SENSOR CLOUD

The proposed risk assessment framework for WSNs in a sensor cloud will determine the likelihood and the impact of the attacks on a WSN. The likelihood of attacks is influenced by factors such as sensor node configuration, topology, and routing measures. Additionally, execution of an attack increases the possibilities of other attacks [35]. These types of interdependencies between the attacks can be modeled using attack graphs. Quantifying these attacks based on CVSS parameters will help us to determine the feasibility of various known attacks. When we merge these attack graphs with the principles of Bayesian networks, we can then estimate the net impact of the feasible attacks on the WSN security parameters. A flow chart summarizing the risk assessment framework is illustrated in Figure 4.2.

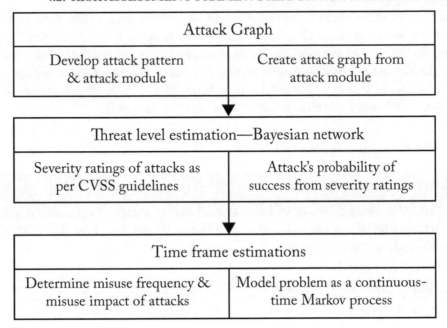

Figure 4.2: Proposed risk assessment framework.

4.2.1   ATTACK GRAPHS FOR WIRELESS SENSOR NETWORKS

Attack graphs, as introduced earlier in this chapter, are graphical structures with a bottom-up assessment approach in which the root node is a critical system asset that we are trying to evaluate based on the identified vulnerabilities in various systems it may interact with. These interconnected systems are depicted as the intermediate and the leaf nodes of the attack graph structure. To generate an attack graph structure, the two pre-requisites are to develop an *attack pattern* and an *attack module* database.

The very first step toward generating an attack graph is to to develop an attack pattern database. It stores data about different attacks pertaining to knowledge such as vulnerabilities they exploit, techniques they use to exploit the identified vulnerabilities, and so forth. This gives us insight about the motivation and goal of an attacker which is then useful in devising security measures that allows us to avert these attacks. An attack on a WSN will tend to exploit one or more of the WSNs security parameters like confidentiality, integrity, and availability. Thus, in the domain of WSNs, we can group the attacks according to the security parameter(s) they tend to exploit. This will help in developing the attack pattern for attacks on a WSN [51].

Definition 4.1   Attack Pattern. An attack pattern is a tuple $P_i = (s_i, \phi)$, where s_i is the attack and $\phi \in$ [Confidentiality, Integrity, Availability] is one of the exploited WSN security parameters.

Attacks on a WSN can be categorized as active or passive attacks [52]. Active attacks, such as sinkhole, are executed to alter the network resources or operations such as routing protocol. Passive attacks, such as eavesdropping, are executed to gather information about the network which can then be used to execute Active attacks. An example instance of classification of active and passive attacks is shown in Table 4.1. An instance of the set of known attacks along with their definition and attack patterns [38, 39, 51] is given in Table 4.2.

Table 4.1: Classification of attacks on a WSN

Types of Security Attacks	
Active Attacks	**Passive Attacks**
Routing Attacks (Spoof, Alter and Replay; Selective Forwarding; Sinkhole; Sybil; Wormhole; HELLO Flood)	Passive information Gathering (Eavesdropping)
Denial of Service (Frequency Jamming)	Traffic Analysis
Fabrication (Node Subversion and Node Malfunction)	Camouflaged Adversaries
Lack of Cooperation (Node Outage)	
Modification (Physical Tampering and Message Corruption)	
Impersonation (Node Replication)	

Further, to understand the cause-consequence relationships between attacks on a WSN, we should be aware of the conditions that are required to execute a particular attack, also known as pre-conditions of an attack, and the consequences of successful execution of an attack, also known as the post-conditions of an attack. If the post-conditions of an attack satisfies the pre-conditions of another attack then these two attacks will have a cause-consequence relationship which is represented as an edge in the attack graph. We develop an attack module that will capture the cause-consequence relationship between the attacks (Table 4.3). In some cases, pre-conditions of an attack may be satisfied by the post-conditions of a single attack, these kinds of attacks will be connected by an *OR* type join in the attack graph (Figure 4.3), whereas, if post-conditions of two or more attacks are required simultaneously to satisfy the pre-conditions of an attack, they are connected by *AND* type joins.

Definition 4.2   Attack Module. An attack module is defined as a tuple, $(P_i, S_{pre}, S_{post}, \epsilon)$, where P_i is the attack pattern, s_{pre} is the pre-conditions required to execute the attack, s_{post} are the post-conditions after the execution of the attack, and ϵ is the join type, $\epsilon \in [OR, AND]$.

Table 4.2: Attack definition and attack pattern

Attack	Definition	Target Security Parameter
Eavesdropping	Listening to communication between the nodes	C
Jamming and DoS	Disrupting network communication by jamming communication frequency or sending additional garbage packets the network cannot handle	A
Node Subversion	Adversary taking over a legitimate sensor node	C; I
Sybil	Adversary node creating false virtual node identities and making them seem like legitimate sensor nodes	C; I
Spoofing	Adversary node pretending to be a legitimate node of a WSN	C; I
Altering; Replay	Continuously changing the route of packets; sending the same packets over and over again	I; A
Wormhole; Sinkhole; Blackhole	Adversary falsely advertising efficient route paths and rerouting traffic from actual paths—all traffic can now pass through the adversary and (s)he may choose to drop all the packets	C; I; A
Selective Forwarding	Once packet traffic gets rerouted through the adversary (s)he may decide to selectively drop some of the packets	C; I; A
Acknowledgment Spoofing	Falsifying acknowledgment during authentication procedure or when legitimate sensor nodes are trying to identify its neighbors	C; I
Node Malfunction	A legitimate sensor node functionally abnormally due to scare resources or a malware running on them	I; A
Node Replication	An adversary creating a rogue sensor node which is based on a legitimate sensor node	C; I
False Data Injection	Adversary introducing garbage packets within the actual packet transmission	I
Node Outage	When a legitimate sensor node is no longer able to function	A
Directed Physical Attack	Physical damage brought unto the sensor nodes	A
Hello Floods	Continuous Hello messages sent to sensor node, which makes them unable to handle any other messages and drains their resources as a result of constantly having to deal with these Hello messages	A
Desynchronization	Sensor nodes constantly trying to re-establish broken communication and not being able to do so	A
Malware Attack	Execution of a malicious code on a legitimate sensor node	C; I; A

Table 4.3: Attack module

Attack	Pre-condition	Post-condition	Join Type
Eavesdropping	Adversary node within physical reach of WSN and unencrypted communication	Acquire unencrypted data	AND
Jamming and DoS	Adversary within the reach of WSN	Disrupt network communication	OR
Node Subversion	Inability to authenticate; running malicious codes	Attacks that require bypassing authentication	OR
Sybil	Node subversion; inability to authenticate	Bypass authentication; breaking of key distribution	AND
Spoofing	Sybil; node subversion	Further degradation of authentication	OR
Altering; Replay	Sybil; spoofing	Loss of energy; loss of integrity	OR
Wormhole; Sinkhole; Blackhole	(Node subversion) Sybil; spoofing	Route control; packet drop—partial or complete; communication disruption	AND-OR
Selective Forwarding	Wormhole; sinkhole	Network more vulnerable to replay and rerouting	AND
Acknowledgment Spoofing	Node subversion; Sybil	Spoofing; wormhole; sinkhole	AND
Node Malfunction	Low snergy of the sensor nodes; execution of malicious codes on the sensor	Erroneous data in the network	OR
Node Replication	Physical tampering	Adversary node legitimate part of WSN	AND
False Data Injection	Adversary node; In communication range of WSN	Presence of garbage data in the WSN; jamming due to more traffic	AND
Node Outage	Severe energy drain; directed physical attack ; blackhole	Network disruption	OR
Directed Physical Attack	Topology discovery	Node outage	OR
Hello Floods	Spoofing; Sybil	Energy drain; breaking route table due to continuous Hello Packet transmissions	OR
Desynchronization	Node outage; Hello floods; replay; reactive jamming	Communication disruption	OR
Malware Attack	Architecture of the sensor nodes; user can execute his/her code	Node subversion	AND

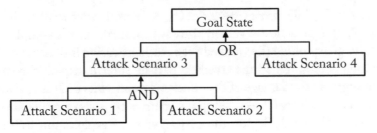

Figure 4.3: An illustration of an attack graph.

Once the attack module has been developed, we can use it to implement attack graphs for each of the WSN security parameters. By doing so, we can visualize the ways in which an attacker might exploit the WSN security parameters. Additionally, the root node of an attack graph is considered to be an attacker's ultimate goal. Hence, exploitation of one of the WSN security parameters—confidentiality, integrity, or availability, will be the root node of an attack graph in our proposed risk assessment framework for a WSN in a sensor cloud. All other nodes in the attack graph will be intermediate states an attacker can take during the course of their attack. In this regard, we assume that an attack state cannot be undone once it has been exploited, preventing any form of backtracking. Therefore, we define an attack graph for a WSN in the sensor cloud using information from the attack pattern and attack module databases as follows.

Definition 4.3   Attack graph. An attack graph is a tuple containing the attributes (s_{root}, S, τ, ϵ), where s_{root} is the goal of the attacker—one of the WSN security parameters. S denotes the complete set of attacks (Table 4.2). τ denotes the set of pre- and post-conditions of all attacks in S. ϵ is the join type, $\epsilon \in$ [OR, AND].

4.2.2   QUANTITATIVE RISK ASSESSMENT BY MODELING ATTACK GRAPHS USING BAYESIAN NETWORKS

An attack graph's nodes maybe assigned with either true or false values implying that an attack state is either successfully executed or not executed at all. However, to better analyze an attack scenario and draw actionable insights from it, we can assign numerical values. An instantiation of such numerical representation can be encapsulated in the *probability of success* of executing an attack s_i, $\Pr(s_i)$. The probability of success of attacks can be derived from their severity ratings and can be estimated by adopting the scoring metric established by NIST's CVSS.

CVSS scores are based on three criteria; base metrics, temporal metrics, and environmental metrics. Base metric is used to express the level of difficulty to exploit a vulnerability. Temporal and environmental metrics are used to express the effect of the network's deployment environment and surroundings in the exploitation of the vulnerability. The base metric is further comprised of *exploitability sub-score* and *impact sub-score*. The exploitability sub-score is comprised of *access vector* (B_AR), *access complexity* (B_AC), and *authentication instances* (B_AU). Access vector (B_AR) depicts how a vulnerability is exploited. Access complexity (B_AC) describes the difficulty of exploiting the vulnerability. Authentication instances (B_AU) states how many authentication is required on the attackers part in order to exploit the vulnerability. The impact sub-score suggests the damage of the attack on parameters such as *confidentiality* (B_C), *integrity* (B_I), and *availability* (B_A). Temporal metrics consists of availability of *exploitability tools and techniques* (T_E). It depicts the current state of exploitation techniques available. *Remediation level* (T_RL) categorizes the type of solutions available to fix the vulnerability. *Report confidence* (T_RC) describes the evidence available on the existence of the vulnerability. The environmental metrics quantify two aspects of impact that depend on the surrounding environment of the organization: (1) the *collateral damage* of the attack (E_CDP); and (2) the *security required* for organizational assets such as confidentiality (B_CR), integrity (B_IR), and availability (B_AR). These metrics are summarized in Table 4.4.

Table 4.4: CVSS metric description

CVSS Metrics	Subcategories	Attributes	Description
Base	Exploitability sub-score	Access Vector	Vulnerability exploitation from within network domain or remotely
		Access Complexity	Dificulty in exploiting vulnerability
		Authentication Instances	Number of authentication measure that needs to be bypassed
	Impact sub-score	Confidentiality	Impact on confidentiality
		Integrity	Impact on integrity
		Availability	Impact on availability
Temporal	Exploitability tools & techniques (T_{E})		Current state of available exploitation techniques
	Remediation level (T_{RL})		Type of solutions available to fix the vulnerability
	Report Confidence (T_{RC})		Evidences available about existence of vulnerability
Environmental	Collateral Damage		Impact of exploited vulnerability on organization's economy
	Security required		Amount of security required for organizational assets like confidentiality, integrity, and availability

WSN attacks have not yet been corroborated by CVSS and as such the base metrics will be evaluated subjectively. We also assume that, although preventive measures for the attacks are available, they are solutions which individual researchers have reported. Hence, based on the definition of *remediation level*, we have considered these solutions as *workaround* fixes. Environmental metrics are context specific varying for different organizations and is constant for any given organization. Houmb's misuse frequency model [46] performs risk level estimations as a conditional probability over the *Misuse frequency* (MF) and *Misuse Impact* (MI) estimates of an attack. MF and MI of an attack helps in depicting the likelihood and impact of an attack respectively taking into account the intrinsic attributes of the attack, network architecture, and security measures used. It is useful in estimating time frames predicting the degradation of organizational assets like confidentiality, integrity, and availability. Hence, to compute the probability of success of attack nodes in our attack graph, we have adopted Houmb's misuse frequency model. MF of an attack is calculated using (4.1)–(4.3) and CVSS parameters specified in Table 4.5:

$$MF_{init} = \left(\frac{1}{3}\right) \sum_{s_i \in S} (B\_\{AR\}, B\_\{AC\}, B\_\{AU\}) \tag{4.1}$$

$$MF_{uFac} = \left(\frac{1}{3}\right) \sum_{s_i \in S} (T\_\{E\}, T\_\{RL\}, T\_\{RC\}) \tag{4.2}$$

$$MF = \left(\frac{1}{2}\right) \sum_{s_i \in S} (MF_{init}, MF_{uFac}). \tag{4.3}$$

Initial misuse frequency, MF_{init} in (4.1) is calculated using exploitability sub-score under the base metrics (Tables 4.4 and 4.5). We normalize the values of B_{AR}, B_{AC}, B_{AU} for the attack under consideration, to keep the final score between 0→1 since the value of MF is a probability and therefore cannot be over 1. The MF of an attack, however, may change over time according to the availability of security solutions and techniques for executing the attacks. These factors are reflected using temporal metrics, computed as MF_{uFac} (4.2). MF_{uFac} is then added to (MF_{init}) and the final misuse frequency (MF) is computed in (4.3). Similar computations are done to calculate MI using the impact sub-score under base metrics and environmental metrics (Table 4.6) [54] and (4.4)–(4.7). Initial MI estimate, MI_{init}, is estimated using impact sub-score of the base metrics in (4.4). This estimate is a vector depicting the effect of an attack on confidentiality, integrity, and availability of a network. MI_{init} is then updated on the basis of the collateral damage potential (E_CDP) in (4.5). The MI estimates are further updated as per the security requirements information in (4.6). Finally, the resulting MI estimate, *MI*, is obtained in (4.7):

Table 4.5: CVSS vectors to calculate MF

CVSS Metrics	CVSS Attributes	Rating	Rating Value
Base	Access vector (B_{AR})	Local (L)	0.395
		Adjacent network (A)	0.646
		Network (N)	1.00
	Attack complexity (B_{AC})	High (H)	0.35
		Medium (M)	0.61
		Low (L)	0.71
	Authentication instances (B_{AU})	Multiple (M)	0.45
		Single (S)	0.56
		None (N)	0.704
Temporal	Exploitability tools & techniques (T_{E})	Unproved (U)	0.85
		Proof-of-Concept (POC)	0.90
		Functional (F)	0.95
		High (H)	1.00
	Remediation level (T_{RL})	Official Fix (OF)	0.87
		Temporary Fix (TF)	0.90
		Workaround (W)	0.95
		Unavailable (U)	1.00
	Report Confidence (T_{RC})	Unconfirmed (UC)	0.90
		Uncorroborative (UR)	0.95
		Confirmed (C)	1.00

$$MI_{init} = [B\_\{C\}, B\_\{I\}, B\_\{A\}] \tag{4.4}$$

$$MI_{CDP} = E\_CDP\,[MI_{init}] \tag{4.5}$$

$$MI_{Env} = [B\_\{CR\}, B\_\{IR\}, B\_\{AR\}] \tag{4.6}$$

$$MI = MI_{CDP} \times MI_{Env}. \tag{4.7}$$

Once we compute the MF, we can estimate an attack's probability of success, $Pr(s_i)$, using (4.8):

$$Pr(s_i) = (1 - \mu)MF_{init} + \mu\,(MF_{uFac}), \tag{4.8}$$

where μ is a constant and is defined as the security administrator's belief of the impact of the security measures on an attack's MF (base metrics) and temporal metrics. It can vary from [0,0.5].

Table 4.6: CVSS vectors to calculate MI

CVSS Metrics	CVSS Attributes	Rating	Rating Value
Base	Confidentiality Impact (B_{C})	None (N)	0.00
		Partial (P)	0.275
		Complete (C)	0.660
	Integrity Impact (B_{I})	None (N)	0.00
		Partial (P)	0.275
		Complete (C)	0.660
	Availability Impact (B_{A})	None (N)	0.00
		Partial (P)	0.275
		Complete (C)	0.660
Environmental	Condentiality Requirement (E_{CR}))	Low (L)	0.50
		Medium (M)	1.00
		High (H)	1.51
	Integrity Requirement (E_{IR})	Low (L)	0.50
		Medium (M)	1.00
		High (H)	1.51
	Availability Requirement (E_{AR})	Low (L)	0.50
		Medium (M)	1.00
		High (H)	1.51
	Collateral Damage Potential (E_CDP)	None (N)	0.00
		Low (L)	0.10
		LowMedium (LM)	0.30
		MediumHigh (MH)	0.40
		High (H)	0.50

If a security administrator is uncertain about the impact of security measures of an attack (like in cases of new or unknown attacks), then the probability of success is based on the base metrics, by taking μ as 0.

After assigning attack graph's nodes with their probability of success, we depict it as a Bayesian network. This will help us to ascertain the non-deterministic nature of the attacks for different network scenario with a reasonable amount of accuracy. A Bayesian attack graph in our framework can be defined as follows.

Definition 4.4   Bayesian Attack Graph.   A Bayesian attack graph is a tuple containing the attributes (S, τ, ϵ, Pr), where S is the complete set of possible attacks on a WSN (Table 4.2), τ, is the set of the pre- and post-condition of all the attacks in S, ϵ is the join type, $\epsilon \in$ [OR, AND], for the attacks in S, and Pr is the set of probability values specifying the success rate of an attack node in a Bayesian attack graph.

For an attack $s_i \in S$, $Pa[s_i]$ denotes the parent set of attack s_i, i.e., if an attack $s_j \in Pa[s_i]$, post condition of attack s_j will lead to pre-condition of attack s_i. As such, attack s_j will be the parent of attack s_i in the attack graph. The probability of each attack's success is captured in a Local Conditional Probability Distribution (LCPD) table. These values are assigned as per the subjective belief of the security administrator regarding their network. LCPD can be defined as follows.

Definition 4.5   Local Conditional Probability Distribution.   For a Bayesian attack graph containing the tuples (S, τ, ϵ, Pr), the local conditional probability distribution function of any $s_i \in$ S is given as $Pr(s_i | Pa[s_i])$ and is defined as:

1. $\epsilon = AND$

$$Pr\left(s_i \mid Pa[s_i]\right) = \begin{cases} 0, \exists s_j \in Pa[s_i] \mid s_i = 0 \\ Pr\left(\bigcap_{s_i=1} s_i\right), otherwise. \end{cases} \tag{4.9}$$

2. $\epsilon = OR$

$$Pr\left(s_i \mid Pa[s_i]\right) = \begin{cases} 0, \forall s_j \in Pa[s_i], s_i = 0 \\ Pr\left(\bigcup_{s_i=1} s_i\right), otherwise. \end{cases} \tag{4.10}$$

Using these evaluations gives us the capability to carry out two forms of qunatitative risk assessments for the WSNs—static risk assessment and dynamic risk assessment. Static risk assessment is typically performed during the pre-deployment phase of the sensor cloud, whereas dynamic risk assessment is carried out during the post-deployment phase.

Static Risk Assessment

The difficulty of executing an attack is given by its probability of success, $Pr(s_i) \forall s_i \in S$, also known as the prior probability. With this set of prior probabilities captured in a node's LCPD, we can compute the unconditional probabilities. Consider the attack scenario described in Figure 4.4.

The probability of attack 1's success is assigned based on the subjective belief of a security administrator. Using the concepts of Bayesian networks, we estimate the probability of success of the remaining nodes in the attack graph. Once we have the probability of success of every node, we calculate the joint probabilities of all the variables in a Bayesian network using the

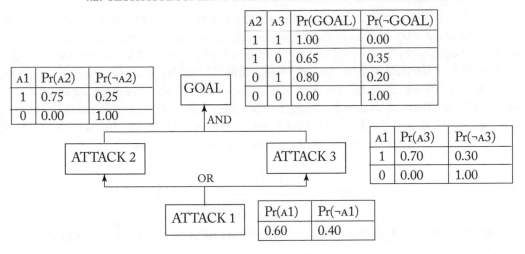

Figure 4.4: An attack scenario depicted by an attack graph represented as a Bayesian network.

chain rule in Eq. (4.11)

$$Pr(s_1, \ldots, s_n) = \prod_{i=1}^{n} Pr(s_i \mid Pa[s_i]).$$ (4.11)

The unconditional probability of the goal state is then computed as the joint probability of all of the nodes effecting the goal state's outcome. Thus, in Figure 4.4:

$$
\begin{aligned}
Pr(Goal) &= Pr(Goal, A3, A2, A1) \\
&= Pr(Goal \mid A3, A2) \times Pr(A2 \mid A1) \times Pr(A3 \mid A1) \\
&= \sum_{A3, A2, A1 \in \{T, F\}} [Pr(Goal \mid A3, A2) \times \\
&\quad Pr(A2 \mid A1) \times Pr(A3 \mid A1) \times Pr(A1)] \\
&= (1.0 \times 0.75 \times 0.70 \times 0.60)_{TTT} + \\
&\quad (0.65 \times 0.75 \times 0.30 \times 0.60)_{TFT} + \\
&\quad (0.80 \times 0.25 \times 0.70 \times 0.60)_{FTT} \\
Pr(Goal) &\approx 0.49.
\end{aligned}
$$

The unconditional probability of each node is computed similarly by considering the sub-graph rooted at that node.

Dynamic Risk Assessment

Static risk assessment is done by assuming non-zero prior probabilities of the attacks. However, once WSNs in a sensor cloud has been deployed, we may observe evidence of certain attacks. The probability of success of that attack node will become one, leading to re-evaluation of risk level estimations. We perform this estimation using the Bayesian inference techniques of forward and backward propagation. Successor of the attack node with probability 1 will be updated by forward propagation. The initial assumptions on all prior probabilities during static risk assessment will be corrected with backward propagation. The updated unconditional probabilities are known as posterior probabilities. Given a set of attacks s_i' for which we have evidence of exploit, the probability of success for those attack nodes is now 1. Thus, we need to determine the probability of success for the attack nodes that are affected by s_i', i.e., the set of $s_j \in \{S - s_i'\}$. We compute the posterior probability, using Bayes theorem as follows:

$$Pr\left(s_j \mid s_i'\right) = \frac{\left[Pr\left(s_i' \mid s_j\right) \times Pr\left(s_j\right)\right]}{Pr\left(s_i'\right)}. \tag{4.12}$$

In Figure 4.4, if we have evidence of the goal state being compromised, we can compute its effect on attack 3 as:

$$Pr(A3 \mid Goal) = \frac{[Pr(Goal \mid A3) \times Pr(A3)]}{Pr(Goal)}$$

Where,

$$Pr(Goal \mid A3) = \sum_{A2 \in \{T,F\}} [Pr(Goal \mid A2, A3 = T)$$
$$\times Pr(A2)]$$
$$= (1.0 \times 0.49)_T + (1.0 \times 0.51)_F$$
$$Pr(Goal \mid A3) = 1.0$$
$$Pr(A3) = 0.42$$
$$Pr(Goal) = 0.49$$

Therefore,

$$Pr(A3 \mid Goal) = 0.85.$$

Hence, the unconditional probability of attack 3 has gone up from 0.49–0.85 in the evidence of an exploit. Similar computations can be done for other nodes.

4.2.3 TIME FRAME ESTIMATIONS

Predicting time frames for degradation of a WSN security parameter is a useful tool to carry out maintenance of an unattended WSN and take precautionary measures before they reach an unrepairable state. The MF and MI estimates of WSN attacks is used to build such a risk level estimation. We model our risk assessment framework as a continuous-time Markov process. Such a model consists of a finite state space E, having n service levels—SL_0 to SL_n. Each service level is a subset in E. We define a service level as follows.

Definition 4.6   Service Level. Service level is a state composing of non-empty sets of attacks $s_k \in S$. Attacks belonging to a service level have equivalent misuse impact.

Attacks are grouped based on their MI on WSN security parameters—confidentiality (C), integrity (I), and availability (A). This gives us the number of service levels in the state transition model. The first service level, SL_0, has no impact on a WSN security parameter, in contrast to the final service level, SL_x which has full impact. The time frame estimation will be a two-step process: (1) develop state transition model from MI estimates: creation of service levels; and (2) compute state transition rates from MF estimates using a rate transition matrix: the probability of transition from a service level with lower impact to that of a service level with higher impact.

4.3   USE CASE SCENARIO DEPICTING THE RISK ASSESSMENT FRAMEWORK

In this section, we illustrate and use case instantiation of the presented risk assessment framework for WSNs present in a sensor cloud elaborating the step presented and discussed in Section 4.2. Let us consider a sensor cloud network consisting of five WSNs. The deployment region of these networks are pre-defined. Three of these networks consist of three sensor nodes, each in addition to the base station and the remaining two consists of four and five sensor nodes, respectively. The sensor cloud is serving three users, tasked to sense three different phenomena— temperature, humidity, and light intensity. Three attack graphs will be generated of each of these five WSNs, one for each of the WSN security parameter—confidentiality, integrity, and availability. For our use case scenario, we are going to illustrate the attack graph and time frame estimations for confidentiality of one of the WSN consisting of three sensor nodes.

4.3.1   ATTACK GRAPH FOR CONFIDENTIALITY

We calculate the probability of success for each attack node in the attack graph to determine the threat level estimations (Section 4.2.2, Eq. (4.11)). This is done assuming that the WSN had no security measures and thus all attacks shown in Table 4.2 are feasible. This assumption helps in illustrating the complete set of attacks on a WSN through an attack graph. In the presence of security measures for a particular attack, the probability of success for that attack node will be zero. For the illustrated attack graph (Figure 4.5), the expected threat level for a particular security parameter was assumed to be 50% which signifies the expected probability of successful exploitation of the goal state, i.e., WSN's confidentiality. The expected threat level indicates the subjective belief of a security administrator about the chances of degradation of their WSN's security parameter due to an attack. In the absence of security measures, there is an equal likelihood of an attack being feasible or otherwise. Although one can argue that absence of security measures should prompt in a higher percentage of expected threat level, but to encompass the uncertainty of an attacker's attack, the assumption of 50% expected threat level is justified for the sake of this illustration. The unconditional probability of success for each attack is given along with nodes of the attack graph. These values are used to compute the net

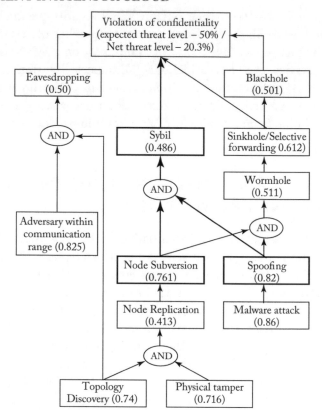

Figure 4.5: Use Case Scenario—attack graph for exploitation of confidentiality of WSN in a sensor cloud.

threat level. They are computed adopting the techniques used to calculate the severity rating for vulnerabilities of wired networks established by CVSS (Section 4.2.2).

We select the attacks from Table 4.2 with confidentiality of a WSN as their attack pattern. These attacks become the attack nodes in the attack graph for confidentiality. Then, we analyze the pre- and post-conditions of each of these attacks from the attack module (Table 4.3), building the logical correlation and depicting it via the attack graph. Once we have created the attack graphs, we need to assign the attack nodes with their probability of success. This is done by evaluating the Base metrics and Temporal metrics (Table 4.5) for the attacks. An instantiation of this evaluation is given in Table 4.7.

Misuse frequency (MF) is then computed using (4.1)–(4.3). Confidentiality of a WSN can be exploited via eavesdropping attack node (Eav) by successful execution of topology discovery

Table 4.7: Use Case Scenario—evaluation of misuse frequency of attacks on WSN

Attack Name	Base_Metrics (B_AR, B_AC, B_AU)	Temporal Metrics (T_E, T_RL, T_RC)	MF_{init}	MF_{uFac}
Eavesdropping	Adj, Low, None	F, W, C	0.686	0.966
Node Subversion	Network, Medium, Single	POC, W, C	0.723	0.95
Sinkhole/Selective Forwarding	Network, Medium, MI	FEE, W, C	0.686	0.966

(TD) and adversary within communication range (ACR) attacks conjunctively (Figure 4.5). We compute the unconditional probability of eavesdropping attack node using (4.9) and MF of these three attack nodes (Eav: 0.82, TD: 0.74, ACR: 0.825). Then, we compute the unconditional probability for Eav using (4.9) as follows:

$$= Pr(Eav)_T * Pr(TD)_T * Pr(ACR)_T$$
$$= (0.82 * 0.825 * 0.74)$$
$$Pr(Eav)_{uncond.} = 0.50.$$

Similarly, the attack node, node subversion (NS), can be exploited via successful execution of either malware attack (MA) or node replication (NR) attack node. The MF of NS, NR, and MA is 0.83, 0.413, and 0.86, respectively. Since we have a disjunctive join in the attack graph, the unconditional probability for node subversion will be computed using (4.10) as follows:

$$\sum_{(NR,MA)\in\{T,F\}} (Pr(NS)_T * Pr(NR) * Pr(MA))$$
$$Pr(NS)_{uncond.} = 0.761.$$

Similar computations are done for other attack nodes. We see from Figure 4.5 that an attacker can exploit confidentiality of a totally unprotected WSN by executing Eavesdropping, Sybil, Blackhole, or Selective forwarding, either individually or in combination, giving them 2^4 attack options. But some of these combinations will not contribute toward the exploitation of confidentiality. For example, Blackhole attack is a successful consequence of Selective Forwarding. If Selective Forwarding does not contribute toward the exploitation of confidentiality, then there will be no contribution from Blackhole. Hence, for an expected threat level of 50%, the computed net threat level for confidentiality will be 20.3%. Since it is very hard to get hold of the information unless the adversary knows the location of the sensor nodes and can closely monitor and capture the traffic. Given the protective measures used, this can be challenging since the adversary must decipher the captured information.

4.3.2   TIME FRAME ESTIMATIONS

Computing the MI of the attacks whose attack pattern is confidentiality (Table 4.2), we have two sets of impact for our use case scenario—0.14 and 0.33 (Table 4.6 and (4.7)), along with service levels SL_0 (fully operational) and SL_x (total degradation). The attacks having an impact of 0.14 are grouped into service level SL_1 and those having an impact of 0.33 are grouped into service level SL_2. Confidentiality degrades and reaches an irreparable state, as we traverse from SL_0 to SL_x. The service levels for WSN security parameters is summarized in Table 4.8.

Table 4.8: Service levels for WSN security parameters using Table 4.2

Service Levels	Attacks	Attack Pattern
SL0(0.0)	-	-
SL1(0.14)	Node subversion, spoofing, node replication, malware attack, wormhole, selective forwarding	C; I
SL2(0.33)	Eavesdropping, sybil, selective forwarding, spoofing, alter/replay, acknowledgment spoofing, node malfunction	C; I
SL3(0.50)	Frequency jamming, denial of service	A
SLx(1.0)	-	-

Computing State Transition Rates

Once the service levels are generated, we compute the rate transition matrix using MF estimates (Table 4.5 and (4.3)). The transition rates for the service level are illustrated in Table 4.9. In this regard, we generally assume that a transition from a higher service level to a lower service level is not feasible to decrease the complexity in computations. Also, we assume that a network cannot reach SL_x directly from SL_0 or SL_1, since SL_0 is a fully operational level with no harmful attacks. Furthermore, execution of attacks in SL_1 will result in a transition to the next service level (SL_2 and not SL_x, since the impact of attacks in SL_1 is lower that of attacks in SL_2). The network would be functioning in SL_0 in the absence of attacks. Execution of an attack belonging to SL_1 causes a traversal from SL_0 to SL_1, and so forth. The transition from SL_1 to SL_2 is dependent on the transition rate for SL_0 to SL_1 and is computed as MF(SL1SL2)| MF(SL0SL1).

The output of rate transition matrix as shown in Table 4.9 is interpreted as follows. Given a time frame of say 30 days, a WSN in absence of security measures will have its confidentiality fully compromised in about 13 days. Since the probability of full compromise, reaching SL_x, is around 0.44, which translates to 13 days (44% of 30 days). We also conclude that in such a WSN there is 66.27% chance that the data will be fully compromised. Integrity, in close co-

Table 4.9: Rate transition matrix for WSN security parameters

	SL0	SL1	SL2	SL3	SLx
SL0	0	$(0.83)_{C,I,A}$	$(0.81)_{C,I}$ $(0.77)_A$	$(0.83)_A$	0
SL1	0	$(0.83)_A$	$(0.66)_{C,I}$ $(0.64)_A$	$(0.69)_A$	0
SL2	0	0	0	$(0.44)_A$	$(0.44)_{C,I}$
SL3	0	0	0	0	$(0.20)_A$
SLx	0	0	0	0	0

relation to confidentiality, will also be lost in 13 days. Since the probability of full compromise, i.e., reaching SL_x is around 0.44, which translates to 13 days (44% of 30 days). This result makes sense because a network cannot maintain its integrity after confidentiality is fully compromised, whereas for availability, there is a 20% chance that a WSN will face communication disruption beyond the point of recovery, since probability of full compromise (SL_x) is around 0.20. Thus, a complete degradation might occur in about 6 days (20% of 30 days) if precautionary measures are not taken to protect the WSN. The rate transition matrix computation for all three WSN security parameters is summarized in Table 4.9.

4.4   DISCUSSIONS

In this section, we discuss some of the essential implications of the risk assessment framework that we have presented in this chapter. Most notably, along the directions of its complexity analysis and scalability when it comes to designing and developing the framework. Second, a discussion related to the differences between a risk assessment framework and an intrusion detection systems and the facet in which these two distinct frameworks can supplement each others functionality.

4.4.1   COMPLEXITY ANALYSIS AND SCALABILITY

In this section, we discuss the complexity involved in designing the presented risk assessment framework and its scalability with respect to large-scale sensor clouds. The initial steps of designing and developing our risk assessment framework involves the creation of a database which contains the information on different WSN attacks regarding their attack patterns, pre- and post-conditions, and JOIN type in the attack graph. Creation of attack graphs requires extracting this information from the database. This process involves considering each attack attribute as a root node and traversing the remaining attack attributes in database and determining if the pre-condition of the attack attribute being considered as the root node matches the post-condition of the remaining attack attributes in the database. We perform this step three times, once for each of the three WSN security parameters. Hence, this process is upper bounded by $O(n^2)$, where n being the number of WSN attacks taken into account. Creation of attack graphs is followed by scoring the attack nodes in the graph with their probability of success which takes a constant amount of time. We then compute the net threat level of the root node using the concepts of the Bayesian networks, computational complexity of which is upper bounded by constant time. Time frame estimation involves the creation of service levels from the Misuse Impacts and then computing the transition matrix from the Misuse frequency values, which is again upper bounded by a constant time. Hence, the total computational complexity of our proposed risk assessment framework is squared.

The framework creates three attack graphs for a given WSN network. As such, if N is the total number of WSNs in the sensor cloud, the total number of attack graphs that will be generated is $3N$. For larger values of N, the number of attack graphs that needs to generated and evaluated increases rapidly. Although this increase is linear and the generation of the attack graphs is not dependent on either the number of sensor nodes or the number of WSNs present in the sensor cloud. Hence, the proposed risk assessment can scale with respect to the increase in number of WSNs in the sensor cloud. However, a challenge lies with the security administrator to draw inferences from net threat level values generated by the attack graph.

4.4.2   RISK ASSESSMENT VS. INTRUSION DETECTION SYSTEMS

In the domain of WSNs in a sensor cloud, Intrusion Detection Systems (IDS) are utilized during the post-deployment phase of the network and presented risk assessment framework is utilized during the pre-deployment phase (especially considering static risk assessment). IDS in general monitors the network traffic or a particular host for evidences of an attack and reports it back to the security administrator if any abnormalities are found. This takes place in real-time, i.e., IDS cannot detect an attack unless the execution of the attack is in progress or has already taken place.

In contrast, the presented risk assessment framework probabilistically determines the overall threat to the network security parameters and helps in deciding the amount and types of security measures that need to be allocated to a WSNs. As such, estimation from the risk assessment framework can be used to determine appropriate placement of an IDS. Additionally, the monitoring aspect of an IDS like system can be used to improve the probabilistic estimation of the risk assessment framework by reporting back evidence of attacks.

Existing IDS for WSNs can detect single or multiple attack pattern(s), but it cannot address the logical relationship between these attack patterns. The presented risk assessment framework can exhibit this logical relationship and have IDS placed detecting a particular attack pattern, identified as critical node in the attack graph, making overall security more efficient and effective. In summary, IDS can trace hosts, report back altercations, monitor network activities, and detect when network is under attack. However, it can neither carry out attack investigation nor identify its placement or make up for weak security measures. All of these drawbacks can be addressed by our risk assessment framework. Thus, performing vulnerability and risk assessment is a prerequisite to developing IDS.

4.5   SUMMARY

In this chapter, we presented a risk assessment framework for WSNs in a sensor cloud environment. We depict the cause-consequence relationship for attacks on WSNs using attack graphs and perform quantitative assessment by representing them as Bayesian networks. Thus, we are able to compute the net threat level to WSN security parameters—confidentiality, integrity, availability. Thereafter, we develop time frames estimating the degradation of these WSN security parameters such that maintenance and repair activities can be scheduled in the largely unattended WSNs before they reach a state of permanent damage.

CHAPTER 5

Secure Aggregation of Data in a Sensor Cloud

5.1 INTRODUCTION

The primary task of any network of sensors is to communicate data to the user. This is usually done via either a gateway or a base station. In a network of wireless sensors this data transmission is done using wireless communication. While wireless communication offer a number of advantages such as ease of deployment, freedom from fixed infrastructure, and flexible routing, wireless transmission tends to consume a large amount of energy. According to [81], the energy consumed in performing computation on 1 bit of data is several orders of magnitude less than in transmitting data for 1 clock cycle. In WSNs and sensor clouds, the communication is often multi-hop. This implies that a packet of data will travel through multiple sensor nodes before reaching the sink. Each node on the routing path receives packets from one of its neighbors and then retransmits it for other nodes. This makes data communication in a WSN a highly energy-consuming task.

One of the ways of reducing the number of data packets in a network is by aggregating data in the network. A number of applications that use WSN data aggregate the received data. Such applications do not need data from individual sensors. For such applications data can either be aggregated by the application itself or by the base station or by the network of sensors themselves. In the rest of the book, we assume that our networks are catering to applications that require aggregate data. Aggregating data as close to the source as possible is beneficial in many aspects, it reduces energy consumption, minimizes end to end delay and bandwidth usage and increases throughput as well as network lifetime. In-network data aggregation is therefore very valuable in WSNs and more so in a large network of sensors such as in a sensor cloud.

Security of data that is sensed and communicated to the base station is very important and may even be critical depending upon the application. A sensor cloud contains a large number of nodes belonging to multiple WSN owners, a subset of which is usually provisioned to a user. The data collected by this set of provisioned nodes may pass through and be aggregated on nodes that are not provisioned to the user. Under such circumstances, it is vital that the data is protected from intermediate nodes. In this chapter, we will take a look at two popular ways to provide security of aggregated data in a sensor cloud.

5.2   RELATED WORK

As mentioned previously, the greatest challenge while working with wireless sensors is the limited battery power available on the sensor nodes. Radio communication consumes a large amount of energy on a sensor and so minimizing communication between nodes is vital in wireless sensor networks. Data aggregation therefore is a topic of considerable interest to researchers in this field. Secure data aggregation schemes are generally classified into two categories: hop-by-hop secure data aggregation schemes and end-to-end secure data aggregation schemes. In hop-by-hop secure data aggregation schemes, security operations are performed at each hop in the routing path, while in an end-to-end scheme security operations are only performed at the source and the sink.

Early secure data aggregation schemes were hop-by-hop schemes. Schemes like [69] mostly dealt with the issue of data confidentiality in the face of a single compromised node. In this scheme, while a parent aggregated the MACs sent by its children, it didn't aggregate the data. Aggregation of data was performed by the grandparent. A compromised node could either corrupt the MACs or the data but not both. Thus, a single compromised node could always be identified. Such algorithms, however, failed when both the parent and the grandparent were compromised. Often when an adversary strikes, it captures a handful of nodes in the attacked region and not just a single node. Schemes tackling the issue of multiple compromised nodes were introduced later, for example the scheme by Chan et al. [70]. This algorithm was resilient to any number of malicious nodes but dealt only with attacks on data integrity. The integrity verification algorithm was distributed through the network which reduced the communication load on certain nodes and increased the time to first node failure. Schemes like SecureDAV [71] and SDAP [72] also provided for data integrity by making use of threshold cryptography and Merkle hash trees. SecureDAV [71] like [70] did not have confidentiality protection. It used threshold cryptography for proving authenticity of the messages and Merkle hash trees for integrity. Like [70], SDAP [72] also aimed to reduce the congestion around the nodes which are placed high up in the hierarchy. To accomplish this, the algorithm logically partitioned the aggregation tree into sub trees of similar sizes called groups. Data integrity was provided by using a commit and attest technique. Hop-by-hop schemes, however, tend to leave the data exposed at the aggregators. A malicious or compromised aggregator can easily compromise the security of a hop-by-hop secure data aggregation scheme. This issue is exacerbated in a sensor cloud environment where aggregators may or may not belong to the same WSN owner and may therefore have added incentives for leaking aggregated data.

End-to-end secure data aggregation schemes were consequently proposed, some of which are discussed in [68, 73–75, 82, 85] and [86]. These schemes use the concept of Homomorphic Encryption which is described in Chapter 2. Homomorphic Encryption is sometimes also called Privacy Homomorphism (PH) in literature. Schemes that use Privacy Homomorphism (PH) for data confidentiality can be classified into two categories: schemes that rely on asymmetric encryption and schemes that use symmetric key encryption-inspired approaches.

The scheme presented in [73] used an Elliptic Curve Cryptography (ECC) based PH for data confidentiality but it didn't provide data integrity. The scheme by Sun et al., on the other hand, presented in [75] provided data confidentiality as well as data integrity. Since then there have been many schemes proposed in the literature that use PH for data confidentiality and one of the many Aggregate Digital Signature Algorithms for data integrity protection. While these schemes are more or less able to achieve the security objectives, they tend to be inefficient when it comes to energy consumption, on account of using energy-hungry asymmetric key cryptography.

Symmetric key-based PH approaches are more efficient in terms of energy. It needs to be noted that, in this text, approaches that don't use classical symmetric key cryptography but do not use asymmetric cryptography either are referred to as symmetric key approaches. Castelluccia et al.'s scheme in [74] used a modular addition-based PH for encryption. The modular addition-based PH had the advantage of simplicity, however, the scheme required the keys to be pre-distributed and did not have a provision for key updates which was a serious security flaw.

Two privacy preserving algorithms, CPDA and SMART, were proposed in [82]. In CPDA, nodes hide their data in a random polynomial and send it to all other nodes in the cluster using pair wise keys. These polynomials are added up and sent to the cluster head, along with a secret. The other algorithm SMART is less communication intensive than CPDA. In SMART, each node in an aggregation tree slices its data into a fixed number of parts and sends each part to a neighboring node, keeping one for itself. Nodes add all of the received slices together and send them to the aggregator along the path of the tree. In CPDA, the aggregate is revealed to the cluster head while in SMART, because positive slices of data are distributed, some amount of information about the data is always leaked. Moreover, the privacy of data depends on key pre-distribution. If an adversary is able to capture a sufficient number of nodes, the privacy in both schemes suffers heavily. The scheme in [85] utilized secret perturbation to address the issue of privacy. This scheme prevented the aggregator from knowing the aggregate, with no privacy loss when a few nodes were compromised. In PASKOS and PASKIS algorithms proposed in [86], each node is assigned a subset of keys from a key ring. In PASKOS, nodes use all of their keys to create hashes, and randomly either add or subtract them from the data. This randomized data and a list indicating whether the hashes were added or subtracted is sent to the aggregator. The authors further improve the algorithm such that even the base station is oblivious to the aggregate in the PASKIS protocol. These protocols protect the aggregate, even from the intermediate nodes. As in [82], however, the privacy depends on key pre-distribution. Moreover, because a list, whose size is the size of the key ring, must be sent by every node, the bandwidth consumption of the protocols is also very high.

Schemes in [82, 85, 86] addressed the issue of privacy of data but ignored the integrity of data. In many cases, the goal of the adversary is to have the base station accept false readings which these schemes fail to protect against.

In this chapter, we present two solutions for the problem of secure data aggregation in sensor clouds. Our first solution based on asymmetric key encryption uses ECC-based encryption scheme for efficient data encryption and an ECC-based aggregate digital signature scheme that is able to aggregate and verify digital signatures. This scheme protects both the confidentiality and integrity of data and is resilient against the false data injection attacks. Our second solution is based on symmetric key approaches and also provides both data confidentiality and data integrity. This solution is also resilient to data injection attacks and, by virtue of being a symmetric key solution, is more energy-efficient than the first solution.

5.3   SECURE HIERARCHICAL DATA AGGREGATION ALGORITHM

The secure hierarchical data aggregation algorithm employs homomorphic encryption and aggregate digital signatures for providing end-to-end confidentiality and data integrity. The encryption algorithm used is Elliptic Curve Elgamal (ECEG), which is the Elgamal encryption algorithm adapted to work on elliptic curves. The digital signature algorithm is a version of the Elliptic Curve Digital Signature Algorithm (ECDSA) modified to allow for signature aggregation. We also designed a secure tree construction algorithm shown in Algorithm 5.1 which favors strong bidirectional links and organizes the nodes in a tree hierarchy. The tree construction algorithm adapts to node disconnection and consists of a reconfiguration procedure shown in Algorithm 5.2 that can be used by disconnected nodes to connect back to the network. This procedure helps in increasing the connectivity of the network under a node capture or denial of service attack.

After the nodes are deployed, they start the tree construction protocol and organize themselves in a tree hierarchy rooted at the base station. Once the tree construction is over, the nodes start sensing data. Each node then generates a reading x. The reading is signed using the aggregate signature algorithm and $SIG(x)$ is generated. The reading is then encrypted using the homomorphic encryption algorithm and the ciphertext $ENC(x)$ is produced. The leaf nodes of the tree send the encrypted data, signature and the public key corresponding to the private key used for signature generation to their parent. After an intermediate node has received data from all its children, it performs a summation of the encrypted readings (SUM-ENC), which is possible because of homomorphic encryption. It also performs the summation of the signature (SUM-SIG) using the aggregate signature algorithm and all the public keys (SUM-PK). SUM-ENC, SUM-SIG, and SUM-PK are then sent to the node's parent. This process is repeated at every node until the data reaches the base station. In the remainder of this section, we discuss the signature and encryption algorithms in more detail

5.3.1   MODIFIED ECDSA SIGNATURE ALGORITHM

The signature algorithm assumes the sensors are preloaded with the appropriate elliptic curve parameters, the base station's public key and a network wide random integer. The random integer is used to compute a new integer k for each round at the sensors. Every sensor therefore generates

---

Algorithm 5.1 Tree Construction Algorithm

Require: Parameters MAX_CHILDREN, NUM_NODES, Secret key k deployed on each node

1: The base station starts by broadcasting a HELLO message, which consists of a random number r.
2: If a sensor which has not yet elected its parent receives a HELLO message, it calculates $HMAC_k(r)$ and sends it in a PARENT REQUEST to the originator of the HELLO message.
3: **if** a PARENT REQUEST is received **then**
4:     Calculate $HMAC_k(r)$ and check whether it matches the HMAC in the packet.
5:     Check whether RSSI value of the request packet is greater than a minimum threshold.
6:     Check that the number of children is less than the MAX_CHILDREN limit.
7:     The number of requests from a particular node is less than the allowed limit.
8:     If the above four checks are satisfied, send an ACCEPTED message otherwise send REJECTED.
9: **end if**
10: Upon receiving an ACCEPTED message a node elects the sender of the ACCEPTED message as its parent and broadcasts a HELLO message.
11: If a sensor has not been able to elect a parent after a certain period of time it invokes the HELP procedure.
12: In case of a disconnection the HELP procedure is invoked by the affected nodes.

Algorithm 5.2 HELP Procedure

1: The sensor invoking the help procedure broadcasts a HELP request with a random number r_1.
2: Nearby sensors who are higher in the hierarchy than the HELP invoking sensor and can accept more children reply to the HELP request with $HMAC_k(r_1)$ and a random number r_2.
3: The sensor calculates $HMAC_k(r_1)$, compares it with the $HMAC_k(r_1)$ in the incoming packets and chooses one of the replying nodes as its parent. It then sends $HMAC_k(r_2)$ to the parent.
4: The parent verifies $HMAC_k(r_2)$ and upon verification accepts the sensor as a child.

---

and uses the same integer k in a given round. At the start of each new data aggregation round, sensors choose their private key z and generate a corresponding public key $Q = zT$ where T is the base point on the elliptic curve. In the original ECDSA algorithm a signature is a tuple (r, s) such that $r = (r(x) \bmod p)$, where $(r(x), r(y)) = kT$, and $s = k^{-1}(h(m) + z * r(x)) \bmod p$. Where h is a secure hash function and p is a sufficiently large prime number. When two signatures $d_1 =$

(r_1, s_1) and $d_2 = (r_2, s_2)$ that were created in the same round on two messages m_1 and m_2 are added, r_1 and r_2 will be equal while s_1 and s_2 can be written as $s_1 = k^{-1}(h(m_1) + z * r(x))$ and $s_2 = k^{-1}(h(m_2) + z * r(x))$. ECDSA is not an aggregate signature scheme because when these two signatures are added $h(m_1)$ and $h(m_2)$ need to be added. Hashing is not homomorphic and, therefore, $h(m_1) + h(m_2) \neq h(m_1 + m_2)$. With the original ECDSA an aggregate signature will not be the same as the signature on the sum of messages. ECDSA can be made additive by a simple modification. If we replace the hash of the message by the message itself in the calculation of (r, s), the signature becomes additive since we are summing up integers when we add s_1 and s_2. The signature (r, s) in the modified signature scheme is $r = (r(x) \mod p)$ and $s = k^{-1}(m + z * r(x)) \mod p$.

The modified ECDSA, which does not use hashing, is vulnerable to the universal forgery attack in which an adversary will be able to get a random message and signature pair accepted by the verifier. An adversary that has captured a node in the network will also be able to attack the modified ECDSA and obtain the private key of any node in the network. When a node transmits the digital signature $(r, s) = [r(x) \mod p, k^{-1}(m + z * r(x)) \mod p]$, an adversary with a compromised node can compute z by using the value of $r(x)$. Integers k and p are known in the network and the value of m can be estimated since in a network of sensors, there is a high probability that nearby nodes will sense similar data. This attack could be thwarted if either the value of m could be made sufficiently random or the value of $r(x)$ could be made unavailable to the attacker. We use the second approach in this solution. We only allow the sensors to send a part of the signature. The sensors only transmit the s component while the r component is generated at the base station. The attacker will now not be able to obtain z or modify the complete signature and hence will not be able to launch a successful universal forgery attack.

5.3.2   EC ELGAMAL ENCRYPTION

The solution further uses the additive homomorphic Elliptic Curve Elgamal Encryption (ECEG) scheme for preserving the confidentiality of sensor data. Before the ECEG scheme could be used to encrypt plaintext data, the data has to be mapped to a point on the elliptic curve. We use a simple homomorphic mapping technique where we multiply the plaintext message m by the base point T, to get the elliptic curve point $M = mT$. Each message m_i maps to a point M_i on the elliptic curve. Once the data has been mapped to a point on the elliptic curve, the ECEG algorithm can proceed. The ECEG algorithm is show in Algorithm 5.3. If the M_is are added, the addition of the elliptic curve points is equivalent to the addition of the plaintext data:

$$
\begin{aligned}
M_1 + M_2 + \ldots + M_n &= map(m_1) + \ldots + map(m_n) \\
&= m_1 T + m_2 T + \ldots + m_n T \\
&= (m_1 + m_2 + \ldots + m_n)T \\
&= \left(\sum m_i \right) T.
\end{aligned}
$$

The final step in the decryption process of the EC Elgamal algorithm is reverse mapping the elliptic curve point back to the plaintext data. Unfortunately, there is no known method that can efficiently perform the reverse mapping of a point to an integer for our mapping function. The reverse mapping function we use is a brute force attack on the elliptic curve point mT given point T. In our secure data aggregation scheme, decryption is only performed at the base station. Our assumption is that the base station is a powerful machine. Therefore, reverse mapping the point using a brute force method aided by the knowledge of the range of sensed data and heuristics such as Pollard's ρ method and the baby-step giant-step method is feasible.

Algorithm 5.3 Elliptic Curve Elgamal

Require: Elliptic curve parameters $D = (q, FR, a, b, T, p, h)$, sensor reading m_i, base station's private key z_e and base station public key $Q_e = z_e T$

   Encryption
1: Map the message m to an elliptic curve point M using a mapping function.
2: Generate a random integer a_i.
3: Calculate $C_1 = a_i T$ and $C_2 = M + a_i Q_e$.
4: $(C_1, C_2) = (a_i T, M + a_i Q_e)$ is the ciphertext.
   Decryption
5: Calculate $(-z_e * C_1)$ and add it to C_2.
6: The decrypted message M is the addition $(-z_e * C_1) + C_2$.

5.4 PRIVACY AND INTEGRITY PRESERVING DATA AGGREGATION (PIP)

This algorithm is based on the Recursive Secret Sharing (RSS) algorithm described by Parakh and Kak in [93]. The scheme was originally proposed for achieving efficient storage relative to the original secret sharing algorithm by Shamir which is described in Chapter 2. Secret sharing algorithms divide data δ into n shares, which are created such that the shares do not reveal any information about the original data and the original data can be recreated only if k shares are combined together where $k < n$. RSS [93] provides a construction where the shares of the data δ can be used to store $k - 2$ additional pieces of information. A node with at least k shares can easily reconstruct all of the $k - 1$ pieces of hidden information. In our algorithm, we use secret sharing for achieving confidentiality, since the shares themselves are completely random and do not reveal any information about the data. We provide a construction in which we can prevent a node which has all the shares from reconstructing and retrieving the hidden data. Our construction preserves the homomorphic property of SSS and RSS which is used to aggregate data in a privacy preserving manner.

The basic idea is to create shares for a given sensor reading using RSS. Each node then sends all of the shares to its parent. Because RSS is based on SSS, it is additively homomorphic and so the parent node will be able to aggregate the shares to aggregate the data. However, because each node is sending all the shares to its parent, it is straightforward for a curious aggregator to regenerate the data and thus subvert the confidentiality. To prevent this, each node scrambles the shares by subtracting modulo prime, a scrambling key shared between each node and the base station from the shares. This scrambling procedure is very lightweight and preserves the homomorphic property of RSS. Homomorphic encryption could be used for this purpose but scrambling is preferred since it consumes negligible energy compared to encryption while still being secure. Finally, to preserve the integrity of the data, the intuition is to have some additional information, along with the shares to help the base station in the detection of any changes in the data. An integrity key is used for this purpose. Each node has its own integrity key which it shares with the base station. This integrity key is embedded in the RSS shares, as RSS allows us to store extra pieces of information in the shares.

5.4.1   THE PIP ALGORITHM

The basic PIP algorithm consists of two sub-algorithms, one for generating the shares from the data and the keys and the other for regenerating data and checking its integrity. The algorithm assumes that the sensor nodes are randomly deployed in a field and they organize themselves into a tree hierarchy using the tree construction algorithm shown in Algorithm 5.1. The base station preloads each sensor with a pseudo-random function (PRF) $f(.)$, a network wide nonce n, three distinct random seeds r_1, r_2, r_3, a network-wide random seed r_4, and the network-wide prime number p. Before every round, a sensor node uses the PRF $f(.)$ on the nonce n, using the random seeds r_1, r_2, r_3, r_4 as keys, to generate the perturbation key η, a scrambling key Ψ and an integrity key $I = \{I', I''\}$, as shown in Algorithm 5.4.

These keys will be used in this round as explained below and will also be used as random seeds for the key generation in the next round. Keep in mind that all the operations are performed mod p. In the basic PIP algorithm, a sensor S_k uses the perturbation key η_k, and the sensor reading δ_k to generate perturbed data $\delta_{\eta k}$. Recursive secret sharing is then used and a linear combination of the integrity key $I'_k + I''_k \delta_{\eta k}$ and data, $\delta_{\eta k}$ are encoded to create the shares $\omega_k^2(3), \omega_k^2(4)$, and $\omega_k^2(5)$. If these shares are sent to the aggregator, then a malicious aggregator would be able to easily replace the data in the shares with corrupt data, keeping the integrity key intact. We use the scrambling key Ψ_k to scramble the shares and prevent this from happening. The scrambled shares λ_k^1, λ_k^2, and λ_k^3 are generated, as shown in the Algorithm 5.4. Each leaf node sends the scrambled shares λ_k^1, λ_k^2, and λ_k^3 to its parent, which aggregates the shares received from all of its children. The aggregate shares $\lambda_{agg}^1, \lambda_{agg}^2$, and λ_{agg}^3 are calculated as $\lambda_{agg}^1 = \sum \lambda^1, \lambda_{agg}^2 = \sum \lambda^2, \lambda_{agg}^3 = \sum \lambda^3$, where the summation is over all children. The base station then uses the sum of the scrambling keys $\sum \Psi$ to unscramble the aggregate shares to obtain $\omega_{agg}^2(3), \omega_{agg}^2(4)$, and $\omega_{agg}^2(5)$. These shares are then used to decode received perturbed data $\omega_{agg}^1(0)$ and the linear combination of integrity keys and the perturbed data $\omega_{agg}^2(0)$. If the equation $\omega_{agg}^2(0) - \omega_{agg}^1(0) I''_k = \sum_{k=1}^n I'_k$ holds, the base station can be sure that the data has

Algorithm 5.4 Share Generation Algorithm

Require: Sensor reading δ_k, PRF $f(.)$, nonce n, random seeds r_1, r_2, r_3, r_4.

1: Generate Perturbation key as $\eta_k = f_{r_1}(n)$, Scrambling key $\Psi_k = f_{r_2}(n)$, Integrity key $I_k = \{I'_k, I''_k\} = \{f_{r_3}(n), f_{r_4}(n)\}$.

2: Update $r_1 = \eta_k, r_2 = \Psi_k, r_3 = I'_k, r_4 = I''_k$.

3: Generate perturbed data $\delta_{\eta k} = \delta_k + \eta_k$.

4: Choose a random number y uniformly from $\mathbb{Z}_p$.

5: Interpolate $(0, \delta_{\eta k})$ and $(1, y)$ to obtain a degree 1 polynomial ω_k^1.

6: Sample the polynomial ω_k^1 at $x = 2, 3$ and interpolate $(0, I'_k + I''_k \delta_{\eta k}), (1, \omega_k^1(2)), (2, \omega_k^1(3))$ to obtain a degree 2 polynomial ω_k^2.

7: Sample ω_k^1 at $x = 3, 4, 5$ to obtain $(3, \omega_k^2(3)), (4, \omega_k^2(4)), (5, \omega_k^2(5))$.

8: $\omega_k^2(3), \omega_k^2(4), \omega_k^2(5)$ are the shares which consist of the sensor reading δ_k and the integrity key I_k of the sensor k.

9: These shares are then scrambled using the scrambling key Ψ_k as follows.

$$\lambda_k^1 = \Psi_k - \omega_k^2(3)$$

$$\lambda_k^2 = \omega_k^2(3) - \omega_k^2(4)$$

$$\lambda_k^3 = \omega_k^2(4) - \omega_k^2(5)$$

10: Final shares are the scrambled shares $\lambda_k^1, \lambda_k^2, \lambda_k^3$.

not been tampered with and is recovered after removing the perturbation. Otherwise, the data has been tampered with and the base station would need to take appropriate steps. Algorithmic details of the share verification and data regeneration process can be seen in Algorithm 5.5.

5.4.2 NUMERICAL EXAMPLE

To understand how the PIP algorithm works, let's take a look at an example. Let us consider a small aggregation three consisting of three nodes, as shown in Figure 5.1. Node A aggregates the data from nodes B, C and itself and sends the aggregate to the base station. Each node generates the scrambling key Ψ, the integrity key $I = \{I', I''\}$ and the perturbation key η. The nodes sense data and using these keys, perturb it and generate scrambled shares. To illustrate the share generation process, let us take node B as an example, which has the scrambling key $\Psi_B = 40$, the integrity key $I_B = 49, 4$, and the perturbation key $\eta_B = 2$. All calculations are done mod 131, which is a network parameter. Node B has data $\delta_B = 22$, to which it adds the perturbation key $\eta_B = 2$ and generates perturbed data $\delta_{\eta B} = 24$. It chooses a random number $y = 7$ and interpolates $(0, 24)$ and $(1, 7)$ to generate the polynomial $\omega_B^1(x) = 114x + 24$. This polynomial is sampled at $x = 2$ and 3 to obtain the points $(2, 121), (3, 104)$. Next, we compute $I'_B + I''_B \delta_{\eta B} = 14$ and interpolate $(0, 14), (1, 121)$, and $(2, 104)$ to generate the polynomial

---

Algorithm 5.5 Data Regeneration and Integrity Check

Require: PRF $f(.)$, nonce n, random seeds r_1, r_2, r_3, r_4.

1: B.S generates the scrambling, perturbation and the integrity keys for all the sensors and computes $\sum_{k=1}^{n} \Psi_k$, $\sum_{k=1}^{n} \eta_k$ and $\sum_{k=1}^{n} I'_k$.

2: Root node unscrambles the shares using $\sum_{k=1}^{n} \Psi_k$ as follows.

$$\omega_{agg}^2(3) = \sum_{k=1}^{n} \Psi_k - \lambda_{agg}^1$$

$$\omega_{agg}^2(4) = \omega_{agg}^2(3) - \lambda_{agg}^2$$

$$\omega_{agg}^2(5) = \omega_{agg}^2(4) - \lambda_{agg}^3$$

3: Interpolate $(3, \omega_{agg}^2(3))$, $(4, \omega_{agg}^2(4))$ and $(5, \omega_{agg}^2(5))$ to obtain a degree 2 polynomial ω_{agg}^2.

4: Sample ω_{agg}^2 at $x = 1, 2$.

5: Interpolate $(2, \omega_{agg}^2(1))$, $(3, \omega_{agg}^2(2))$ to obtain a degree 1 polynomial ω_{agg}^1.

6: Calculate $\omega_{agg}^1(0)$ and $\omega_{agg}^2(0)$.

7: **if** $\omega_{agg}^2(0) - \omega_{agg}^1(0)I''_k == \sum_{k=1}^{n} I'_k$ **then**
    Accept the aggregate as $\omega_{agg}^1(0) - \sum_{k=1}^{n} \eta_k$.

8: **else**
    Aggregate is corrupt.

9: **end if**

$\omega_B^2(x) = 69x^2 + 38x + 14$. We sample this polynomial at $x = 3, 4, 5$ to generate the recursive shares $94, 91$, and 95. We then use the scrambling key $\Psi_B = 40$ to obtain the final scrambled shares $\lambda_B^1 = 77, \lambda_B^2 = 3$, and $\lambda_B^3 = 127$. Node B sends its scrambled shares to node A, as does node C. Node A, being the aggregator, aggregates its own shares with the shares it receives to obtain the aggregate shares $\sum \lambda_{agg}^1 = 28, \sum \lambda_{agg}^2 = 78, \sum \lambda_{agg}^3 = 9$. The aggregate shares are then sent to the base station. Because the keying material of all the nodes is shared with the base station, the base station can calculate $\sum \Psi, \sum I', I''$, and $\sum \eta$ which, in this case, are $88, 39, 4$, and 6, respectively. The base station then unscrambles the aggregate shares using $\sum \Psi$. The original shares after unscrambling come out to be $60, 113$, and 104. These shares are interpolated to obtain the polynomial $\omega_{agg}^2(x) = 100x^2 + 8x + 53$. The polynomial $\omega_{agg}^2(x) = 100x^2 + 8x + 53$ is then sampled at $x = 1, 2$ to obtain points $(1, 30)$ and $(2, 76)$, respectively. Points $(2, 30)$ and $(3, 76)$ are then interpolated, and the polynomial $\omega_{agg}^1(x) = 46x + 69$ is obtained. The base station then checks whether the equation $\omega_{agg}^2(0) - \omega_{agg}^1(0)I''_k = \sum_{k=1}^{n} I'_k$ holds. Since we can see that the equation is satisfied in the example, $\omega_{agg}^1(0) = 69$ is taken as the sum of the perturbed data $\sum \delta_\eta$ from which $\sum \delta$ can be derived as $\sum \delta = \omega_{agg}^1(0) - \sum \eta$.

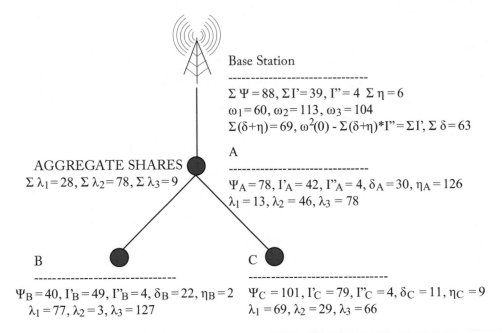

Figure 5.1: Data aggregation using PIP.

5.5 SUMMARY

For a sensor cloud to function, it is critical that a secure data aggregation scheme be used to collect data from the sensors. The secure data aggregation must provide data confidentiality and data integrity in the presence of corrupt aggregators that may be interested in exposing the data being aggregated on them or injecting false data in the aggregate. In this chapter, we have presented two secure data aggregation algorithms which achieve that objective using different cryptography constructs. The first algorithm uses elliptic curve cryptography-based homomorphic encryption and additive digital signature algorithms while the second one uses recursive secret sharing scheme along with scrambling and random perturbation. In a sensor cloud environment, which consists of wireless sensors, it is imperative that our algorithms consume as less energy as possible and have low latency. With that in mind, we recommend symmetric key-based approaches over asymmetric key approaches since they are generally energy efficient. Care must be taken, however, in ensuring that the symmetric key approaches used are proven to achieve the security objectives. The PIP algorithm we presented can be proven, for example, to be semantically secure, as shown in [107].

CHAPTER 6

Access Control of Aggregated Data in Sensor Clouds

6.1 INTRODUCTION

Sensor clouds share some of the same security issues as WSNs and introduce some new ones owing to multiple users and multiple WSN owners [96]. The security issue we consider in this chapter is access control of sensor data. Sensor clouds are distributed systems which generate data in real time and on-demand. The data is aggregated in-network by using algorithms like PIP which means that the access control mechanism should support secure data aggregation. To aggravate the issue, in sensor clouds, the network topology is not known in advance since the nodes are provisioned and de-provisioned on-demand for each user. Moreover, a sensor cloud composed of multiple WSNs may be owned by different users each of which may place certain restrictions on who can and cannot access the data being generated by their sensors. This possibility is unique to sensor clouds and, thus, what is required is an access control scheme for sensor clouds where only authorized users are able to access data from the sensors. The access control scheme should be able to work in an environment where the data is aggregated in network and the topology of sensor nodes can vary for each user and each query. This access control scheme should also be able to integrate the WSN owner permissions on who can access data from their sensor nodes.

A naive solution for this is to use Access Control Lists (ACLs) in wireless sensors where each sensor verifies the access allowed for each user by validating them against its ACL. This solution, however, is not scalable and not suitable in a dynamic environment where users can leave and new users can join. A solution which has been proposed in [97] and [98] is to encrypt data such that only the authorized users are able to decrypt it. The solutions in [97] and [98], however, have considered only a partially distributed system. The system model in these works does not consider an in-network data aggregation scenario. User access control schemes have generally been designed for standalone sensors or smaller networks and do not take large networks into consideration. In large networks, data is generally aggregated in-network during data collection.

In this chapter, we will discuss an attribute-based user access control scheme, which takes into account the fact that data is usually aggregated in WSNs to save energy and bandwidth. The scheme also considers that each WSN in the sensor cloud may impose authorization restriction on the data usage. In this scheme, data instead of being stored on a node is collected in response to the user's query and is aggregated during the process. Furthermore, an extension is discussed where the network owner or the sensor network itself can change the authorization level of the data during run time under special conditions. The sensors under this scheme are able to differentiate between different users who might have the same query, attributes and authorization level. This is of particular help in a commercial system which needs to bill the users according to their usage and so needs to differentiate between users running the same query.

To summarize, the scheme discussed in this chapter provides the following:

- a fully distributed, fine-grained access control algorithm for aggregated data in sensor clouds;

- an algorithm which considers runtime modification of authorizations; and

- integration of the access control scheme with a secure data aggregation scheme.

6.2   RELATED WORK

User access control in sensor networks has received considerable attention in recent years. Using Access Control Lists (ACLs) would seem like a good first step for providing access control where each sensor verifies the access allowed for each user by validating them against its ACL. This, however, is not scalable and suitable in sensor cloud environment where users can leave and new users can join. A simple approach to access control in sensor networks was presented in [99]. In this approach, a user receives a symmetric key K_i for each sensor node SN_i it wants to access. This key is generated by a trusted authority and is based on the user's credentials. The drawback of this scheme is that the user needs to know the identities of all the sensor nodes it wants to access. The scheme also does not support data aggregation, which implies that to access data from multiple nodes, the user will need to individually communicate and get authenticated with each node. A similar approach was presented in the context of service oriented architecture in [100]. This approach has the same drawbacks as that in [99]. In addition to this, in these approaches the access is not fine grained; users are grouped by roles and all users in a particular role have the same access.

In [101], an ECC-based approach was used. The trusted authority issues an ACL to the user based on the user's credentials and issues a public private key pair along with a certificate. This certificate is presented to a local sensor node when the user requires data. The sensor node verifies the certificate before granting access rights. To access data from a remote node, the user needs to get endorsements from k local nodes. These endorsements are verified by the remote nodes before granting access. In this approach, the user need not know the identities of the nodes in advance but like the previous scheme, this scheme too does not support data aggregation.

More fine-grained access control approaches have been presented in [97, 102] and [98]. These solutions are based on the attribute based encryption (ABE) methods. The system model in [102] is different from the conventional system models used for WSNs. In this paper, the access is controlled by the wireless device itself rather than being co-ordinated by a trusted authority. The scheme uses CP-ABE [103] for access control of data, where the access policy is embedded in the ciphertext itself. However, it is computationally expensive when compared to other ABE methods. In [97], each attribute of the data is associated with a public key component. An access tree based on the attributes of required data is created for a given user. The access tree is then used to generate the private key which is provided to the user. The user then provides the access tree to the sensor and the sensor provides data to the user according to the access tree. The data is encrypted such that it can only be decrypted with a private key generated based on that access tree. This scheme was further enhanced in [98] to include multiple base station each of which controls a set of attributes. The concept of ABE was first introduced in [104]. Goyal et al. then proposed key policy-based ABE in [105]. The ABE schemes were proposed for wired systems, where the data is stored at a server and the access is provided according to the attributes held by various users. The schemes in [97] and [98] are direct adaptations of KP-ABE in WSNs and do not take the distributed nature of sensor clouds into account. In the system model presented in [97] and [98], users access data either directly from sensor nodes or from a storage node which collects data from other sensors. While this is a feasible model, the heterogeneity limits the flexibility of the network. In this chapter, a dynamic attribute-based access control mechanism for sensor clouds is presented which works under the general assumption of a homogeneous network.

6.3   MODELS

6.3.1   SYSTEM MODEL

The sensor cloud system as illustrated in Figure 6.1 consists of three parties: the Users ($\mathcal{U}$), the Sensor Cloud Administrator (SCA), and the Sensor Nodes (SN). The sensor nodes are grouped in the form of individual WSNs which are maintained by the WSN owners. To access data from the sensors, a user first contacts the SCA with a query over the attributes of the data it wants to access. It then receives a secret key over these attributes from the SCA. This secret key is then used to access data from the wireless sensors, which has been encrypted based on the said attributes.

6.3.2   ADVERSARY MODEL

We assume that the adversary is capable of capturing a certain percentage of sensor nodes. The adversary has the following goals:

- to try to get access to the data, for which it does not have the secret key;

- to try to access data for which the user does not have authorization; and

- to tamper with the keys and data meant for other users, so as to disrupt the protocol.

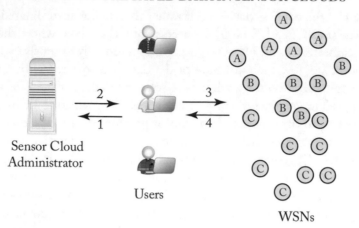

Figure 6.1: System model.

6.4   ACCESS CONTROL POLICY

To obtain fine-grained access control of data, a tree-based access structure was proposed in [105]. The basic idea behind such an access control structure is that each data item can be described by certain attributes attached to it. In the case of sensor clouds, the data being generated by the sensors has certain descriptive attributes such as the type of sensor which generated the data, type of mote on which the sensor is mounted, sensor's location, sensor's owner etc. When users require data from the sensor cloud, they can formulate complex queries based on these attributes to exactly specify the kind of data they want. An example of such a query can be a user requesting data which is either of sensor type T_1 or T_2, from sensors which belong to any two of the three sensor owners—O_1, O_2, and O_3—and which are deployed in region, R_1 . Such a query can be represented by a tree, called an access tree $\mathcal{T}$, as shown in Figure 6.2. In the access tree $\mathcal{T}$, leaf nodes are associated with the attributes while each non-leaf node represents a threshold gate. The logic gate AND can be represented by a $2 - of - 2$ threshold gate, while an OR can be represented by a $1 - of - n$ threshold gate.

Such an access tree is very expressive and can be used to define a number of queries. In a distributed environment, however, there are certain types of queries which cannot be expressed by this access tree. For example, if in the previous query, the user requests data from both regions R_1 and R_2, then an access tree as shown in Figure 6.3 would be formed. This access tree may work in the centralized models of [97] and [98], since the data has already been collected and stored on a single node. In a distributed model, however, no node would satisfy the criteria of being in regions R_1 and R_2 simultaneously and, thus, no data would be returned.

To be able to satisfy queries such as the one described above, partial satisfaction of access trees is required. To accomplish this a modification of the original access tree is required. To perform the modification we first classify the attributes into two types.

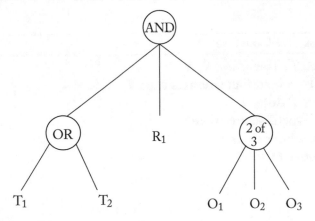

Figure 6.2: Example access tree.

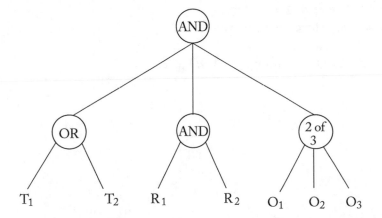

Figure 6.3: Unacceptable access tree.

- **Type A** attributes are those attributes whose multiple instances can be satisfied by a single sensor, e.g., sensor type attribute with instances humidity and temperature can be satisfied by a node which has both humidity and temperature sensors mounted on it.

- **Type B** attributes are the attributes which can only be satisfied one instance at a time, e.g., region attribute. A sensor can only be in one region at a given point of time.

The access tree of Figure 6.3 is modified in such a manner that each top-level sub-tree can be satisfied individually by a sensor. To convert the access tree of Figure 6.3 to an acceptable and partially satisfiable form, the procedure shown in Algorithm 6.6 can be used. The partially satisfiable form of the access tree is shown in Figure 6.4. Figure 6.4 has two top-level subtrees, each of which can be individually satisfied by a sensor which is either in region R_1 or R_2.

Algorithm 6.6 Access Tree Conversion

Require: Access Tree $\mathcal{T}$, Tree arrays X, B
 1: Remove all the Type B attribute subtrees from $\mathcal{T}$.
 2: **for** each subtree $\in T$ **do**
 3:   **if** subtree is a Type B subtree **then**
 4:     Remove the subtree from $\mathcal{T}$.
 5:     Add the subtree to B.
 6:   **end if**
 7: **end for**
 8: Add $\mathcal{T}$ to X.
 9: **for** each subtree $\in B$ **do**
10:   Create as many copies of the access trees in X as there are attributes in the subtree.
11:   Concatenate one attribute to each copy.
12:   Store the concatenated access tree back in X.
13: **end for**
14: Connect all the new trees through an AND gate.

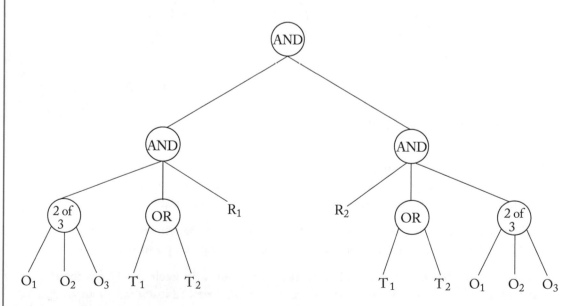

Figure 6.4: Acceptable access tree.

6.5   OVERVIEW OF THE SCHEME

As mentioned above, the system consists of a set of users $\mathcal{U}$, SCA, and the sensor nodes SN. The scheme described in this chapter is derived from KP-ABE that was discussed in Chapter 2. In the setup phase, the SCA generates the public and the master keys. The public key is assumed to be known by the users as well as the sensors. When a user $\mathcal{U}_j$ wants to access data from the sensor cloud, it approaches SCA, with its query on the data attributes. SCA uses the query to create an access tree $\mathcal{T}_j$ and a secret key SK_j over $\mathcal{T}_j$ and $\mathcal{U}_j$'s temporary identity. $\mathcal{T}_j$, SK_j and the public and private components of the user's temporary identity are then given to the user.

A user may contact any of the sensor nodes for data. We designate the node which receives the user request as the Gateway Node (GN_j). After receiving the query, in the form of the access tree $\mathcal{T}_j$, GN_j floods the network with this query, creating a query tree $\mathcal{Q}_j$. Each node in, $\mathcal{Q}_j$, which satisfies the access tree $\mathcal{T}_j$, then generates a random session key and encrypts this key using the key policy attribute-based encryption. The encrypted random session keys of the query tree $\mathcal{Q}_j$, are aggregated and sent to the user via GN_j. The user can then decrypts the aggregated session key, since it has the secret key SK_j. Once the user is able to compute the aggregated session key, it can derive the symmetric keys to be used in data aggregation scheme being used by the system. In this chapter, we will use PIP as a representative secure data aggregation protocol for our explanation. It needs to be noted though that any secure data aggregation protocol which uses a summation of individual sensor keys at the base station and provides confidentiality and integrity can be used.

6.6   ACCESS CONTROL SCHEME

6.6.1   SYSTEM SETUP

The SCA defines the set of all attributes $\mathcal{A}$. For each attribute $i \in \mathcal{A}$, a number t_i is chosen uniformly at random from $\mathbb{Z}_q^*$. A number y is also chosen randomly from $\mathbb{Z}_q^*$. SCA then generates a Paillier public-private key pair (K_{pub}, K_{pr}) and creates the public key as

$$PK = \left(G_1, P, T_1 = t_1 P, T_2 = t_2 P, \ldots, T_{|\mathcal{A}|} = t_{|\mathcal{A}|} P, K_{pub}, Y = e(P, P)^y\right).$$

The master secret key is

$$MK = \left(t_1, t_2, \ldots, t_{|\mathcal{A}|}, y\right).$$

Each sensor is then loaded with, the public key PK, a PRF $f(.)$ and a nonce x.

6.6.2   ACCESS CONTROL SECRET KEY GENERATION

As illustrated in Figure 6.1, when a user $\mathcal{U}_j$ wants to collect data from the sensor cloud, it first contacts the SCA for access. Along with its request, $\mathcal{U}_j$ also provides a query based on the attributes of the data it wants to access. The SCA then creates an access tree $\mathcal{T}_j$, based on the user's query and the attributes, such as the one in Figure 6.2. The access tree is then augmented by ANDing a new node, "*ID*" to the root node as shown in Figure 6.5. This makes sure that the secret key is always bound to an *ID*. This would help in easy revocation at a later time. To create a unique identity attribute for the user, the *SCA* then randomly generates a previously unused number t_j from $\mathbb{Z}_q^*$ and its public component $t_j P$. Once the access tree $\mathcal{T}_j$ is created, *SCA* proceeds to generate the secret key SK_j as follows. Starting from the root node, *SCA* constructs a random polynomial u_x of degree $d_x + 1$ for each node x in $\mathcal{T}_j$, where d_x is the degree of that node. For the root node r it sets $u_r(0) = y$ and chooses the rest of the points randomly. For all other nodes it sets $u_x(0) = u_{parent(x)}(index(x))$, where $parent(x)$ denotes the parent of node x and $index(x)$ returns an enumeration on the children of the parent of node x. All other points are chosen randomly. The secret key SK is then defined as

$$SK_j = \left(D_k = \frac{u_k(0)}{t_k} P, k \in \gamma \right),$$

where, γ is the set of leaf nodes in $\mathcal{T}_j$ including the node *ID* and D_k is calculated for all the leaf nodes k in $\mathcal{T}_j$. *SCA* then gives the secret key SK_j, the access tree $\mathcal{T}_j$, nonce b, and T_j to the user $\mathcal{U}_j$.

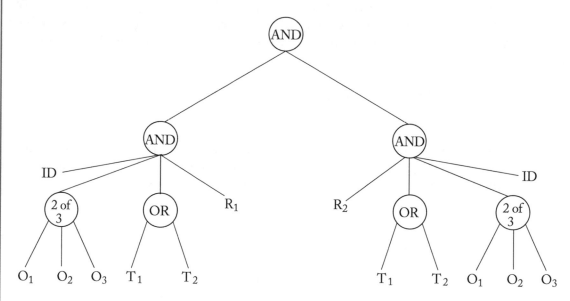

Figure 6.5: Access tree augmented with ID.

6.6.3 DATA AGGREGATION KEY GENERATION

When a user $\mathcal{U}_j$ contacts GN_j for data, it provides the query as the tuple $< \mathcal{U}_j, \mathcal{T}_j, T_j, r >$, where $\mathcal{T}_j$ is the access tree for the query, T_j is the public component of the user's identity and r is a user generated pseudo-random number. GN_j floods this query in the sensor cloud which results in a query tree $\mathcal{Q}_j$. Each node i whose attributes satisfy the access tree $\mathcal{T}_j$, uses the PRF $f(.)$ on the nonce b with r as the key to generate a value $s_i = f_r(b) \in \mathbb{Z}_q$. Each node then generates a partial session key $p_i \in G_T$. which is then encrypted using Paillier encryption to obtain C_{p_i}. This encrypted partial session key C_{p_i} is then again encrypted as

$$E = \left(E' = C_{p_i} Y^{s_i}, \{E_k = s_i T_k\}, k \in \gamma \right). \tag{6.1}$$

Leaf nodes then send the encrypted partial session key to their parent, where it is aggregated as

$$E = \left(E' = \prod C_{p_i} Y^{s_i}, \left\{E_k = \sum s_i T_k\right\}, k \in \gamma \right),$$

which is equal to

$$E = \left(E' = \prod C_{p_i} Y^{\sum s_i}, \left\{E_k = T_k \sum s_i\right\}, k \in \gamma \right).$$

The final aggregate ciphertext of the query tree $\mathcal{Q}_j$ at the gateway node is

$$E_{\mathcal{Q}_j} = \left(E' = CY^s, \{E_k = sT_k\}, k \in \gamma \right), \tag{6.2}$$

where $s = \sum\limits_{i \in \mathcal{Q}_j} s_i$ and $C = \prod\limits_{i \in \mathcal{Q}_j} C_{p_i}$ is the Paillier encrypted aggregate session key.

6.6.4 DATA AGGREGATION KEY ESTABLISHMENT

The ciphertext in Eq. (6.2) which consists of the encrypted session key is given to the user, who proceeds to decrypt the ciphertext. The decryption process works in a bottom up manner, starting from the leaf nodes of the access tree $\mathcal{Q}_j$. For each leaf node k, the value F_k is calculated using:

$$F_k = \begin{cases} e(D_k, E_k) = e(P, P)^{su_k(0)} & if, k \in \gamma_j \\ \bot & otherwise. \end{cases}$$

For each non leaf node z, if z is a d out of n gate and if more than $n - d$ children return $\bot$ then F_z is $\bot$ otherwise F_z is calculated as

$$\begin{aligned} F_z &= \prod_{i \in S_z} F_i^{\delta_i(0)} \\ &= \prod_{i \in S_z} e(P, P)^{su_i(0)\delta_i(0)} \\ &= e(P, P)^{su_z(0)}, \end{aligned}$$

where S_z denotes the set of node $z's$ children and $\delta_i(0)$ denotes the Lagrange coefficient. The user $\mathcal{U}_j$ then computes F_z for the root node, which returns $e(P, P)^{ys} = Y^s$, if the operation is successful. Since E' in the received ciphertext in Eq. (6.2) is CY^s, the user simply divides E' by Y^s to obtain C, which is the aggregate Paillier cipertext of the partial session keys. C can be decrypted to obtain the aggregate of the session keys, S (We would call it the session key). Once the user U_j obtains the session key S, it is used to derive keys for the secure data aggregation algorithm. The secure data aggregation algorithm, PIP, described in Chapter 5 uses three keys: perturbation key η, integrity key (I', I''), and scrambling key Ψ. The keys η, I', and Ψ can be easily derived once S is known while I'' is assumed to be public knowledge. One way to derive the keys can be by using simple modular functions, as shown in Eqs. (6.3), (6.4), and (6.5). k_1, k_2, k_3 here are assumed to be known to the user as well as the sensors:

$$\eta = S \bmod k_1, \tag{6.3}$$
$$I' = S \bmod k_2, \tag{6.4}$$
$$\Psi = S \bmod k_3. \tag{6.5}$$

6.6.5   DATA AGGREGATION

The sensors use the same formulae in Eqs. (6.3), (6.4), and (6.5) on their individual partial session keys to derive their individual data aggregation keys. The user will have the sum of the perturbation, integrity, and scrambling keys of the nodes in the query tree Q_j. Nodes then encrypt data using the PIP scheme and send the ciphertext to their parent. The encrypted data gets routed along the edges of Q_j, being aggregated at each intermediate node. Finally, the encrypted aggregated data is communicated to the user by the gateway node. The user can then use its data aggregation keys to retrieve the data.

6.7   DISCUSSION

In a sensor cloud, there can be multiple users who request the same attribute-based data. In previous approaches, no distinction was made between various users if their query was same. In a sensor cloud though, it is essential to keep track of each user's usage so that they can be billed accordingly. To account for this, the access tree was augmented with an identity element that makes each tree unique even though the attribute based query may be the same. In the access control secret key generation phase, the secret key is generated using the private key component of this identity element along with the other attributes in the access tree. Since the sensors would need the public component of this identity to be able to encrypt data for it, the SCA gives the public component of the identity T_j to the user, which is passed on to the sensors along with the query. The sensors use T_j as the unique identifier for the user's query for billing purposes.

In the data aggregation key generation phase, the sensors generate a partial session key and encrypt it using KP-ABE [105]. The partial session key is encrypted with Paillier encryption [106] to make use of its homomorphic property. In Eq. (6.1), we see that for two sensors SN_1 and SN_2, $E'_1 = p_1 Y^{s_1}$ and $E'_2 = p_2 Y^{s_2}$, respectively. Both the random numbers s_1 and s_2 and the partial session keys p_1 and p_2 need to be aggregated. Since p_i is a Paillier encrypted ciphertext of the partial session key, E_1' and E_2' can simply be multiplied to obtain $E' = p_1 p_2 Y^{s_1+s_2}$. This adds the random numbers s_1 and s_2 as the exponents of Y and multiplies the Paillier ciphertexts p_1 and p_2. From the homomorphic property of Paillier encryption, we know that when the ciphertexts are multiplied, the plaintext is added up. Thus, both the random numbers and the partial session keys are aggregated.

It needs to be noted that Paillier encryption is not being used for security in this scheme. The security instead is provided by the Decisional Bilinear Diffie–Hellman (DBDH) Assumption. Paillier encryption is used for its ability to perform summation on the plaintexts when ciphertexts are multiplied. This introduces the overhead of an extra Paillier encryption during key establishment. This overhead can be reduced by pre-computing n Paillier encrypted elements in G_T and deploying them on the sensors. In the data aggregation key generation phase, the sensors then randomly choose $k \leq n$ elements and multiply them together to generate one random Paillier encrypted partial key. The number of unique keys which can be generated with this method are

$$\text{Number of unique keys } (|p|) = \frac{n!}{k!(n-k)!}.$$

Figure 6.6 shows the relationship between n and k for 80 and 128 bit symmetric key equivalent security.

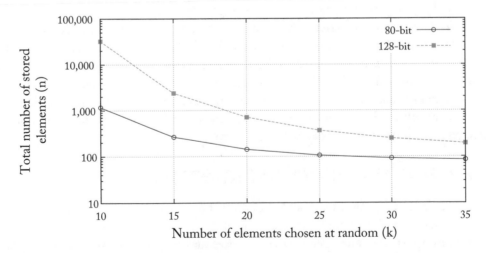

Figure 6.6: Relationship between number of elements chosen (k) and number of unique elements stored (n).

6.8 REVOCATION OF USERS

In FDAC [97], user revocation is handled by updating the master key secret y embedded in the user secret key SK. To accomplish this, the authority SCA includes an additional updatable component in the user secret key SK and a corresponding component in the public key PK. To revoke a user, a new master secret y' and the corresponding public component $Y' = e(P, P)^{y'}$ is generated. The new public component Y' is broadcasted to the sensor nodes, while the differential between the new secret components is broadcasted to all users except the one whose access is to be revoked. The selective broadcasting is done by using ciphertext based attribute based encryption (CP-ABE) [103].

The scheme presented in this chapter also uses an additional secret key and corresponding public key component for user revocation. In Section 6.6.3, it was explained how the access tree is augmented with an ID node and a unique secret key component t_j and public key component T_j is computed for each user U_j. The public component T_j is used by the sensors when encrypting data for the user U_j. Each sensor generates $s_i T_j$ for the user U_j during the data aggregation key generation. $s_i T_j$ is then used in key establishment. This added component hence serves two purposes. First it binds the key to the user. The data aggregation key would be established only if the user possesses the corresponding secret key component t_j. This can be used to keep track of individual data usage of the users and is important in commercial systems like sensor clouds. Second, it helps in easy revocation of users. To revoke a user, the SCA simply has to broadcast U_j in the sensor cloud. Once the sensor nodes receive U_j, they simply stop sending $s_i T_j$ in further data aggregation key generations. This makes sure that the revoked user U_j would not be able to generate data aggregation keys required to decrypt sensor data.

Our user revocation scheme only incurs one broadcast in the sensor network and the maintenance of a list of revoked users. On the other hand, the revocation scheme in FDAC [97] requires one broadcast in the sensor network and one selective broadcast to the users. The selective broadcast is done using CP-ABE [103] which requires computational overhead at the SCA and for each user U_j for every revocation. Moreover, the users may be mobile and they may be online as well as offline. If the selective broadcast fails to reach a user, that user will not be able to receive data from the sensor network in the future.

6.9   MODIFYING ACCESS AT RUNTIME

Data generated by a sensor network may hold different degrees of importance at different times. Sensor data, which under normal circumstances is allowed to be accessed by everyone, may become more important under special circumstances and may require special privileges to be accessed. Two examples below illustrate the point.

Example 1: A network owner has deployed A WSN in a building which keeps track of the number of people coming in and out of the building. This information is public and may be accessed by anyone. In case of a planned public performance event in the building, however, the network owner may escalate the authorization level of this information so that only authorized personnel may access the information about the event.

Example 2: A network owner has deployed seismic sensors in a wide field. This data is public and may be accessed to study the seismic activity of the area. In case the sensors, sensing a sudden short high-intensity shock wave, it may by an indication of an explosion or an earth quake. The sensors themselves escalate the authorization level in this case, to avoid panic in public.

6.9.1   ENCRYPTION SCHEME FOR MODIFYING ACCESS AT RUNTIME

The network owner's or sensor's control on the access of data can be built by introducing authorization in the key establishment phase. We specify authorization by authorization levels which are assumed to be hierarchical in nature. In this scheme, authorization level l is higher than authorization level $l + 1$, such that $AL_0 > AL_1 > \ldots > AL_{n-1}$. A user which has an authorization level of AL_l can access data which belongs to any authorization level AL_m, where $m \geq l$ but cannot access data which belongs to authorization levels, where $m < l$. The hierarchical authorization levels are modeled by a one-way key chain:

$$K_0 \to K_1 \to K_2 \to \ldots \to K_{n-1}, \text{ where } K_l = h\left(K_{l-1}\right).$$

In the key chain, K_0 is the root key and all other keys can be derived by repeated application of the hash function $h(.)$ on the root key. A key can derive other keys in the direction of the hash chain but not the opposite direction. The key chain is such, that data meant for an authorization level AL_l can only be decrypted if the user has a corresponding key K_m, where $m \leq l$. The root key K_0, corresponds to the absolute authorization over all data.

To allow run time modification of access the scheme is modified as follows. Additional public key components $h_l = Y^{K_l}$ are created for each of the authorization keys K_l, such that the public key becomes

$$PK = \left(G_1, P, T_1, T_2, \ldots, T_{|A|}, Y, h_l \text{ for every key } K_l\right).$$

The secret key for each authorization level are the key K_is which are only given to authorized people for that authorization level. To encrypt a message for a certain authorization level l, the public key component of the authorization level is raised to the random number s_i and multiplied with the partial session key. The ciphertext generated by each node in Eq. (6.1) then becomes

$$E = \left(E' = C_{p_i} h_l^{s_i}, \{E_k = s_i T_k\}, k \in \gamma\right). \tag{6.6}$$

The aggregate ciphertext at the gateway node would be

$$E_{Q_j} = \left(E' = CY^{sK_l}, \{E_k = sT_k\}, k \in \gamma\right). \tag{6.7}$$

As discussed before, the user $\mathcal{U}_j$ can use its secret key SK_j on E_{Q_j} to obtain F_z for the root node, which is Y^s. If the user has the correct authorization key, it can calculate $(Y^s)^{K_i} = Y^{sK_i}$, which can be used to retrieve C from (6.7). Once C is obtained, the user can proceed to generate the data aggregation keys.

6.9.2 PROTOCOL FOR MODIFYING ACCESS AT RUNTIME

Once the symmetric keys for data aggregation have been established using the session key, the data collection becomes operational. To modify access to the data, after the data collection is operational, nodes follow the below protocol.

Nodes $SN_1, \ldots, SN_N$ are collecting data and encrypting it securely using keys derived from partial session keys $p_1, \ldots, p_N$. The authorization level is AL_l and the aggregate session key is P. Let's assume that some event $\mathcal{E}$ occurs and in response to $\mathcal{E}$, nodes $SN_k, \ldots, SN_{k+k'}$ escalate their authorization level to $AL_{l'}, l' < l$. The nodes $SN_k, \ldots, SN_{k+k'}$ now generate new partial session keys p', which will be encrypted with the escalated authorization level key $h_{l'}$ as in Eq. (6.6). This ciphertext is sent to the user along the query tree $\mathcal{Q}$. However, to enable the user to keep decrypting the data it receives from the rest of the nodes, the sensor nodes also send their previous partial key p encrypted with the old authorization level. Both the new and the old partial keys of the nodes $N_k, \ldots, N_{k+k'}$ which have escalated their authorization level are aggregated and sent to the user. The user decrypts the old aggregate partial keys $\sum p$ and generates the new session key as $P - \sum p$. This new session key is then used to derive the new scrambling, integrity, and perturbation keys for continuing to receive aggregate data from the rest of the nodes at the old authorization level. $\sum p'$ can only be decrypted by the user if it has the key for the escalated authorization level. Thus, if the user does not have the required authorization to decrypt data from a set of nodes it continues to only receive data from the rest of the nodes. If the user has the private key for the new authorization level, then it can decrypt $\sum p'$ and generate new scrambling, integrity and the perturbation keys. The nodes after escalating the access level now encrypt their data with the new partial keys. A tag tg is attached to this encrypted data to indicate the new authorization level. Encrypted data which has the same tag is aggregated. Thus, the user now receives two aggregates of data, one encrypted with keys derived from the new session key and one with keys derived from the old session key. The user can decrypt one or both the aggregates depending upon the authorization it has.

6.10 SECURITY ANALYSIS

As discussed in the adversary model, the adversary (which is also a malicious user) has the following three goals:

- to try to get access to the data, for which it does not have the secret key;

- to try to access data for which the user does not have authorization; and

- to tamper with the keys and data meant for other users, so as to disrupt the protocol.

To achieve these goals, the adversary can collude with other users and compromise some sensor nodes. We show in this section that the scheme presented in this chapter prevents the adversary from accomplishing these goals in the presence of user collusion and node compromise.

Each node in the scheme encrypts a randomly generated partial session key using KP-ABE. KP-ABE is secure under the Decisional Bilinear Diffie-Hellman (DBDH) Assumption (The security proof can be found in [105]). With KP-ABE, even though the users collude, they cannot decrypt data which has not been encrypted for their individual secret keys. In Eq. (6.7), the session key is encrypted as $Ce(P, P)^{sK_l}$. Assuming, the user has the correct authorization key K_l, in order to recover C, the user will have to calculate $e(P, P)^s$, which it cannot calculate unless it has the correct secret key. The adversary may also compromise sensor nodes. Because each sensor node generates an encrypted partial session key C_{p_i} individually, compromising sensor nodes does not give the adversary any advantage too.

Authorization in the scheme is provided in the form of private keys K_l, for authorization level AL_l. Since the authorization keys are in the form of a one way hash chain, a lower authorization level's key can be derived from a higher authorization key but not vice versa. That is, if the underlying hash function is secure. For a particular authorization level AL_l, the sensor nodes encrypt the session key as $Ce(P, P)^{sK_l}$. Nodes with the required authorization level can calculate $e(P, P)^s$ and then $(e(P, P)^s)^{K_l}$ to compute C and hence the session key S. The user who does not have the correct authorization will not be able to compute $(e(P, P)^s)^{K_l}$ and hence cannot derive the correct session key to access data.

An adversary can compromise sensor nodes, which can corrupt the ciphertext $Ce(P, P)^{sK_l}$ to disrupt the key establishment process. If the user receives corrupted ciphertext, the key generation process will still go through and the user will generate incorrect data aggregation keys. This is where PIP algorithm's integrity preserving nature comes in. Since the user has derived incorrect data aggregation keys, PIP will raise an alarm when the user tries to decrypt sensor data using the incorrect keys. When this happens the user can inform the SCA, which will take appropriate steps.

6.11   SUMMARY

In this chapter, we presented a user access control scheme for sensor clouds. This access control scheme considers large sensor networks where sensor nodes collaborate and aggregate data in the network to save energy and bandwidth. The scheme also provides the opportunity to the network owners to modify the access control policies at run time and provides an efficient revocation strategy. Finally, this scheme is also able to distinguish between different users with the same query, which is very important for sensor cloud applications from an account management and billing point of view. We conclude with the observation that although the scheme is designed for sensor networks, it can also be adopted in other settings such as wired and mobile networks, where the goal is to provide access control in a collaborative and data aggregation scenario.

CHAPTER 7

Efficient and Secure Code Dissemination in Sensor Clouds

7.1 INTRODUCTION

Sensors in a sensor cloud are provisioned to users on-demand and de-provisioned when not in use. When sensors are provisioned to a new user, the code running on them may need to be either updated or changed completely. This happens many times in a sensor's lifetime in a sensor cloud. A sensor cloud is typically very large in size, which makes manual updates to each sensor node impractical. A more plausible alternative is to disseminate the code wirelessly in the network from the base station. This code is routed through multi-hop transmission to the provisioned sensors. Sensor nodes receive the code packet-by-packet and then rebuild the code image, once all of the code has been received. In wireless code dissemination, code images are communicated via the wireless channel which is inherently in-secure and prone to attacks from adversaries. A secure code dissemination technique enables the code dissemination to be confidential and also protected against malicious code injection attacks.

A large amount of work has been done to reduce the amount of code to be transmitted from the base station to different sensor nodes [109–113]. These efforts, however, have been focused on traditional WSNs that support only one application. In such networks, the code updates happen infrequently. Most often, these updates are minor so most of the code remains unchanged. Several schemes, such as [109, 111] and [113], created a difference script between the old and the updated code. The base station disseminates the script rather than the entire code. Doing so not only reduces the number of packets but also saves energy along the forwarding nodes. In a sensor cloud, however, clusters of nodes are provisioned dynamically to the user to support several applications on-demand, [95, 96]. Dynamic provisioning implies that the code on the wireless sensors is changed entirely as a new application is installed. The difference script mechanism cannot be applied in this scenario because the script itself would be the size of the code. Thus, there exists a need for an efficient code dissemination scheme that is well suited for a sensor cloud scenario. Efficiency of code dissemination is an especially important issue in sensor clouds because the frequency of code change is high to support different applications. High frequency of code change implies that the sensors spend a good deal of their energy on forwarding and installing new code. Any reduction in the amount of total code transferred therefore gets multiplied by high frequency resulting in great reduction in energy consumption. Moreover, clusters of sensors in a sensor cloud are dynamically provisioned to users, which means that at any given point in time various clusters can be working for various users. In such a scenario, the security of the code (in terms of confidentiality and integrity) also becomes very important. Code disseminated from the base station will inevitably be forwarded by many sensors on its way to its destination cluster. Code confidentiality is thus a critical pre-requisite since the code may be carrying keying material which must be protected against eavesdropping. Another pre-requisite is the integrity of code, which will make sure that an adversary has not injected malicious code packets during the code dissemination process. To summarize, there is a need for an efficient code dissemination schemes which is well suited to a dynamically provisioned sensor cloud scenario. The scheme needs to minimize the number of code packets transmitted and should provide confidentiality and protect the integrity of the disseminated code. In this chapter, we present an efficient and secure code dissemination scheme for sensor cloud. The scheme is based on storing the common code between two TinyOS applications on the sensors themselves. The technique makes use of similarity in the compiled application code in TinyOS. It also uses proxy re-encryption (PRE) and Bloom filters to additionally provide confidentiality and integrity of the code. In this chapter, Our contributions are as follows:

- a code dissemination algorithm which reduces the total number of packets sent from the base station to a cluster of sensors in a sensor cloud scenario and

- a security mechanism which provides confidentiality and integrity of code packets while they are disseminated in a sensor cloud.

7.2   RELATED WORK

Deluge [114] was one of the first wireless code dissemination protocols developed especially for sensor networks. It provided an efficient way to reprogram motes wirelessly: it divides the code image into fixed size pages and then divides each page into packets with a size that is network dependent. Because Deluge [114] was created for traditional WSNs, it advertises new code images using an epidemic protocol. Nodes can then request for individual pages and new code images by listening to the advertisement packets. Although Deluge [114] tried to reduce the wireless traffic, it did not take into account the similarities between various code updates.

Reijers et al. [113] proposed an approach that addressed incremental code updates. This approach took a UNIX **diff** like approach for determining the difference between two versions of the same code. They introduced commands such as *insert, copy, repair, repair dbl*, and *patch list* to generate an edit script. Instead of wirelessly sending the complete new code, the base station would only transmit the edit script. The wireless sensors would then transform their code image according to the edit script to generate the new version of the code. The authors, however, only discussed the encoding mechanism of the code image without any code distribution algorithm.

Moreover, the edit script was platform dependent. A platform independent incremental code update algorithm was described in [111]. This algorithm divided a program image into small, fixed-size blocks and calculated the hash of each block. The same was done for the new code image and a difference script was created by comparing the hashes of the code images. The difference script consisted of copy and download commands where copy meant the wireless sensor could just copy the block from the old code image and download meant the block contained changes and thus had to be downloaded from the base station.

Approaches in [111] and [113], however, work particularly for small code changes. Any code change that produces an address shift results in an extremely inefficient script. To overcome this, [112] introduced slop regions around functions. Functions are allowed to grow into the slop regions. Thus, small changes in code do not produce address shifts. If the function grows bigger than the allowed slop region, it is moved to an area with a bigger slop region and linked to its previous location. This approach, however, wastes a large amount of space on the sensors. Additionally, the efficiency of code dissemination depends on the amount of memory available to be sacrificed. QDiff [109] presented an optimized patch creation technique. The size of the patch created in QDiff [109] is small compared to other schemes. The algorithm works on the ELF file level and, hence, is platform independent. QDiff [109] uses slop region to maintain similarity between two versions of the code. If no slop region exists, the new code is moved to the end of the file. A high level of similarity between the codes at the ELF file level ensures a small patch size. Moreover, the patch can be directly applied in the RAM, eliminating the need for a reboot and thus saving a large amount of energy.

All of these approaches targeted the traditional WSN model in which the changes between successive versions of the application code are small. On the contrary, in a sensor cloud, a code update implies that the entire application must be changed. As a result, the application code needs to be completely updated. Code dissemination in a sensor cloud like scenario was first discussed in [115]. The authors discussed a code dissemination algorithm for multi-application WSNs, where various groups of sensors support different WSN applications. This algorithm is based on the idea that WSN applications share a lot of common code. This common code can be disseminated from other sensors in the network instead of disseminating everything from the base station. The authors presented the idea and demonstrated the effectiveness through simulations. They did not, however, offer implementation specific details, such as the handling of code shifts and the introduction of new global variables. Instead, the focus was on adaptive buffer management for such a code dissemination algorithm. The algorithm presented in this chapter exploits the same idea that many applications in WSNs may share a large amount of code. Unlike [115], however, we provide a set of algorithms for implementing this in a secure fashion in the context of a sensor cloud.

Another important issue in a sensor cloud scenario is security. Seluge [116] was one of the early attempts to build a secure code dissemination algorithm. Seluge [116] aimed to tackle the code image integrity and various DoS attacks on Deluge [114]. For each code image, it hashes the code packets of the last page and concatenates the hashes with the packets from the previous page. This process is performed recursively until all of the packets from the first page are hashed. The hashes from the first page are then used to create a hash tree, the root of which is signed by the base station's private key. Page 0 is then constructed which consists of the information needed to verify the root hash. In the end, a signature packet is constructed which consists of the root hash, the meta data about the code, and the signature over the root hash. Secure Deluge [117] also applies similar techniques for secure code dissemination. Neither [116] nor [117], however, offer confidentiality of code. Confidentiality of code has been discussed in [118]. In [118], all packets are considered to be in a sequence. The hash of a packet is generated in the same way as in [116]. The last packet, however, is concatenated with an L-byte nonce. The first L-bytes from the hash of a packet are also used as the key for encrypting the packet. The whole sequence of packets is thus encrypted. The hash of the first packet is then used to construct a cipher puzzle signed by the base station. The eavesdropper in [118] is an outsider. The confidentiality of the code is protected against such an adversary, while the nodes in the network are easily able to decrypt the encrypted code. In a sensor cloud scenario, on the other hand, the adversary is inside the network since different clusters of sensors may belong to different users. The confidentiality of the code must be protected against such sensors belonging to other users. This problem is exacerbated by the fact that we also want to store commonly used code on the sensors in the sensor cloud. We propose using proxy re-encryption [119, 120] to solve this problem. Proxy re-encryption is discussed in Section 2.2.5 and we present our algorithm in Section 7.7.3.

7.3 SYSTEM MODEL AND ASSUMPTIONS

The sensor cloud consists of a large number of wireless sensors. We consider that a clustering algorithm such as [121] has been run and the sensors have been grouped into clusters. These sensors are provisioned to the users in terms of clusters. At any given point in time, the sensor cloud may have many users, each holding one or more clusters of sensors (Figure 7.1). Sensors in a cluster provisioned to a particular user collect data for that user. They may, however, act as forwarding nodes for other clusters, for transferring data and code. In other works [109–113], the code updation occurred on the scale of the WSN. In our model, on the other hand, the code change occurs at the cluster scale. The code is updated when either individual users install new applications in their sensor clusters or a cluster is provisioned to a new user and a new application is installed. We assume that a routing structure is in place, using which the base station can route the code to any particular cluster. Each cluster has a cluster key (CK), known to the cluster members and the base station. The adversary in our model lies inside the network. The sensors in the clusters, provisioned to other users are assumed to be curious and may want to eavesdrop on the code being transferred. The sensors storing common code may want to inject malicious code by modifying the code they store and making other sensors accept this modified code.

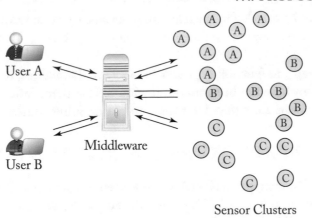

User A

Middleware

User B

Sensor Clusters

Figure 7.1: System model.

7.4   PROPOSED APPROACH

Our approach is based on the observation that the executable code which runs on the wireless sensors consists entirely of subroutines and objects. These subroutines and objects have a one-to-one correspondence with the functions and global variables in a high level language such as nesC in TinyOS. Many of these functions and global variables are common across a number of wireless sensor applications. In a sensor cloud environment which has a number of WSN applications running simultaneously in various clusters of sensors, many applications may share parts of the same code. Some applications may have the same security code, while others may share the same routing subroutines. Still others may share the sensing code, and so on. All applications also share the same operating system code. The basic idea, therefore, is to first identify the commonly used functions and global variables across all of the given applications. These functions and global variables are then distributed throughout the network so that every sensor node probabilistically stores a few of them. When the code on a cluster of sensors needs to be changed, the base station first determines which of the functions and objects can the sensors request from the other sensors in the network. Only the part of code not already present in the network is sent from the base station; the remaining code is requested from nearby sensors. Moreover, since the provisioning of sensors in a sensor clouds is always in the form of clusters, the security-related tasks such as decryption, authentication, and verification can be distributed among the sensors to reduce energy consumption. The security challenges that we will tackle are as follows.

- Because the functions are stored on sensors, a request for a specific function will leak information about the code. To avoid such information leaks, the functions must be stored encrypted on the sensors. Encryption, however, presents a problem because the encryption keys will need to be revealed to the requesting sensors. Once the requesting sensors know the encryption keys, they can send spurious requests and retrieve all of the encrypted code.

- When sensors reply to function requests with encrypted functions, they need to make sure that the functions have not been tampered with. This authentication needs to be done as soon as the functions are received to thwart energy draining attacks.

7.5 THE EC-BBS PROXY RE-ENCRYPTION SCHEME

The concept of Proxy Re-encryption (PRE) was described in Chapter 2. BBS was the first PRE scheme and was introduced in [120]. We present the elliptic curve version of the BBS scheme in Algorithm 7.7 and call it the EC-BBS scheme. The symbols used in the algorithm as well as in the rest of the chapter are described in Table 7.1.

The EC-BBS as well as the Symmetric Re-encryption [119] described in Chapter 2 have the often undesired property of bi-directionality, where it is relatively easy to derive $RK_{j \to i}$ from $RK_{i \to j}$. As we will see in this chapter, however, in the algorithm we present, the knowledge of $RK_{j \to i}$ is not useful for an attacker in gaining any information about the encrypted code. While any PRE scheme can be used in our algorithm, we consider EC-BBS and SRE because of their

---

Algorithm 7.7 EC-BBS Algorithm

Require: Elliptic curve parameters, base point T, message m.

    KeyGen

1: Generate a random integer a. The secret and public key pair then is $(a, a * T)$.

    Encryption

2: Map the message m to an elliptic curve point M using a mapping function.

3: Generate a random integer r.

4: Calculate $C_1 = M + r * T$ and $C_2 = a * r * T$.

5: $(C_1, C_2) = (M + r * T, a * r * T)$ is the ciphertext.

    Decryption

6: Calculate $M' = C_1 - (a^{-1} * C_2)$.

7: Reverse map point M' to message m'.

    Re-encryption KeyGen

8: For nodes A and B, with the secret and public key pair $(a, a * T)$ and $(b, b * T)$, the re-encryption key $RK_{A \to B}$ can be generated as b/a.

    Re-encryption

9: Node A's ciphertext: $C_a = (M + r * T, a * r * T)$.

10: Calculate $C_b = (M + r * T, a * r * T * RK_{A \to B})$.

11: Node B's ciphertext: $C_b = (M + r * T, b * r * T)$.

Table 7.1: Symbols used in the algorithm

Symbol	Description
FID	Unique identifier for each function
CFL	List of common functions
A_i	Authentication key
$h()$	Hash function
$g()$	A pseudo-random function (PRF)
$f()$	A function in the code (not a mathematical function)
$f_j()$	Function with identier j
Q_i	EC-BBS public key
K_i	EC-BBS secret key corresponding to Q_i
$E_{Q_0}(f_j())$	Function $f_j()$, with FID j encrypted using key Q_0
$RK_{i \to j}$	Re-encryption key to re-encrypt a ciphertext encrypted with key Q_i to one encrypted with key Q_j
RK_i	Re-encryption key to re-encrypt a ciphertext encrypted with key Q_0 to one encrypted with key Q_i
CK	A cluster's group key
n	Nonce
S_n	Sesion key to encrypt code packets for a session created using the nonce n
P	Number of pages of code
N	Number of packets in each page
SN	Denotes the sensor network
SN_i	i^{-th} node in the sensor network

energy efficiency. We use EC-BBS to present the algorithm in Section 7.7. To use SRE, a single symmetric key can be used to replace the secret and the public keys in the algorithm.

7.6   DETECTING COMMON FUNCTIONS

We assume that the base station has a tentative list of sample applications that may be used in the sensor cloud in future. It must be noted that not all the applications are needed to be known beforehand. Rather, a small sample size would be sufficient to detect the common functions across the applications. We follow a procedure similar to Qdiff [109] to detect common functions in the applications. The ELF files were dumped using the *msp430-objdump* utility for TelosB and the *avr-objdump* utility for Mica2 platforms. Bauhaus-toolkit [123] was then used to compare the C files generated from the nesC code of the various applications. Bauhaus-toolkit [123] has a clone detection utility that can detect Type I and Type II clones in different applications, at the source code level. Type I clones are fragments of code which are exactly identical while Type II clones are copies which are structurally identical but may have the identifiers changed. In the work presented here we only consider Type I clones. The Type I clones found in the C code by the clone detection utility of the Bauhaus toolkit [123] can be further divided into two different types at the ELF file level.

Definition 7.1   *We define **Type 1a** clones as the true Type 1 clones, where the two codes are exactly the same and they may or may not have been shifted in memory.*

Definition 7.2   *Some Type 1 clones may also contain calls to functions and global variables which have shifted in memory. We define such clones, which have calls to functions and refer global variables, which have shifted in memory as **Type 1b** clones.*

Once the common functions are detected, the SCA base station rearranges the application code in the ELF file in a way which facilitates further use of common functions. The reordering of the functions is done so that the placement of the common functions is consistent across all the applications. Beginning with the Type 1a functions, the base station places the common functions at the end of the *.text* section. This is in contrast to QDiff [109], which places the new code at the end of the *.text* section. The code will now grow toward the beginning of the file. After all the Type 1a functions are moved, Type 1b functions are moved in the same manner. This results in the reordering of other functions and changes in function references. The base station then fixes the changes in function references throughout the code. We place the common functions at the end of the *.text* section and not at the beginning to avoid situations in which the base station receives applications with large *.data* and *.bss* sections. This will move the beginning of the *.text* section further down. In such a case, the common functions placed in the beginning of the *.text* section of a smaller sample application cannot be used. Therefore, it is necessary to ensure that there are no common functions present at the beginning of the *.text* section. This process of detecting common functions and rearranging them in the application code is illustrated in Algorithm 7.8 and Figure 7.2.

7.7   PROPOSED ALGORITHM

7.7.1   PRE-DEPLOYMENT PHASE

The pre-deployment phase consists of two parts: code processing and crypto pre-processing. The common function detection as explained in the previous section is performed in the code-processing part. The base station begins by going through all the sample applications, assigning

Algorithm 7.8 Rearrange Application Code (RAC)

1: Create a List L of Type I clones in the application codes.
2: Mark the elements of L as Ta or Tb, depending on whether they are Type 1a or Type 1b.
3: **for** each i in L **do**
4:     **if** Ta **then**
5:        Move to the last possible location.
6:     **end if**
7: **end for**
8: **for** each i in L **do**
9:     **if** Tb **then**
10:       Move to the last possible location.
11:     **end if**
12: **end for**
13: Reorder the remaining functions.
14: Fix the function calls.

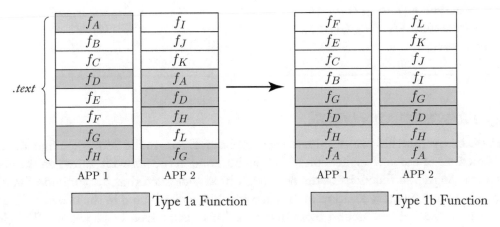

Figure 7.2: Rearranging application code.

a unique identifier, FID, to each new function it encounters. The FIDs and the corresponding functions are stored in a table called the function table. Once all the functions and the common functions are identified, the base station uses Algorithm 7.8 to rearrange the code for each application. In the crypto pre-processing phase, the base station creates a one-way hash chain, $A_0, A_1, \ldots, A_t$ which we call the authentication hash chain. t is taken to be sufficiently large so as to cover the entire lifetime of the sensor cloud operations. The hash chain has the following rules.

1. $A_i = h(A_{i+1})$.

2. A_0 is the root of the chain, which is obtained by applying the hash function $h()$, t times on A_t.

For each sensor, some FIDs are randomly selected. The base station then generates a random secret key K_0 and the corresponding elliptic curve public key $Q_0 = K_0 * T$. The functions are encrypted with this key Q_0 using the proxy re-encryption scheme EC-BBS. Pairs of FID and the associated encrypted function $(j, E_{Q_0}(f_j()))$, authentication key A_0, hash function $h()$, and a pseudo random function (PRF) $g()$ are pre-deployed on the sensors. This process is shown in Algorithm 7.9.

Algorithm 7.9 Pre-Deployment

1: Associate each function with a unique FID and build the function table.
2: Call Alg RAC to create L and rearrange application codes.
3: **for** each i in *SN* **do**
4:     Randomly select k FIDs from L.
5:     **for** each j in k **do**
6:         Encrypt $f_j()$ using Q_0.
7:         Store tuple $(j, E_{Q_0}(f_j()))$ on SN_i.
8:     **end for**
9: **end for**

7.7.2   PRE-DISSEMINATION

When the base station has to disseminate a new application code in the network, it first identifies those functions of the new application which can be found in the network, stored on the nodes. It then rearranges the functions of the new application code such that the common functions reside in the same memory location as the code which was distributed in the network. The rest of the code is then placed around these functions. This can be seen in Figure 7.3. The global variables of this new application are also arranged according to the common functions' need in the *.data* and *.bss* sections. The base station then creates a list called the common functions list (CFL) which is in the form of FIDs along with the size of the functions and their memory location in the compiled code.

Before disseminating the code, the base station generates a random secret key K_i and the corresponding elliptic curve public key $Q_i = K_i * T$ for the i^{-th} iteration of code change. It then computes the re-encryption key RK_i from K_0 and K_i as K_i/K_0. A pre-dissemination packet is constructed which consists of an HMAC of hash of the re-encryption key concatenated with hash of the CFL, i.e., $HMAC_{A_i}(h(RK_i)||h(CFL))$. The key A_i used to generate the HMAC is the next key in the authentication key chain. The HMAC is then disseminated in the network just

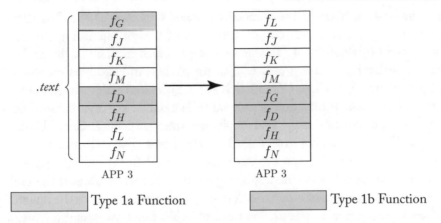

Type 1a Function   Type 1b Function

Figure 7.3: Rearranging application code for a new application.

prior to the code. This is a broadcast packet and all the nodes in the network save the contents of this packet to authenticate the CFL and the re-encryption key at a later stage. This process is illustrated in Algorithm 7.10.

Algorithm 7.10 Pre-Dissemination

1: Create CFL.
2: For the i-th iteration calculate the re-encryption key RK_i from keys K_i and K_{i-1}.
3: Compute $h(CFL)$ and $HMAC_{A_i}(h(RK_i)||(h(CFL)))$.
4: Disseminate $HMAC_{A_i}(h(K_i)||(h(CFL)))$ in the network before the code is disseminated.

7.7.3 CODE DISSEMINATION

The base station (BS) prepares for code dissemination by creating a Bloom Filter (BF) of an appropriate length. It uses a hash function to hash the common functions on the CFL one-by-one to populate the BF. It then combines the CFL, the BF, the new code, the next re-encryption key (RK_i), and the next decryption key (K_i) together. The BS also creates an index on the code dissemination content to help the nodes recover everything; it appends this index to the contents.

Once the total code dissemination content, as shown in Figure 7.4, is known, it is divided into pages. These pages are further divided into packets. The BS then uses a nonce (n) and the cluster's group key (CK) with a PRF $g()$ to generate a session key, $S_n = g_n(CK)$. This key is used to encrypt the packets and provide confidentiality. Each packet is encrypted individually with the same key. To provide code integrity, we used a process similar to that used by Seluge [116]. For the sake of continuation, we use the same nomenclature as that used by Seluge [116]. We assume there are P pages and that each page has N packets. The pages are denoted as Page 1 to Page P, while the packets for Page i are denoted as $Pkt_{i,1}$ to $Pkt_{i,N}$. Packets in Page P are hashed and the hash of packet i is appended to packet i in page $P - 1$. The packets in Page $P - 1$ then consist of the concatenation of the hashes of the corresponding Page P packet and the original packets of Page $P - 1$. This process is continued until all of the packets of Page 1 are hashed. A Merkle Hash Tree is created over the packets in Page 1, as shown in Figure 7.5; we call this the Vertical Hash Tree (VHT). Seluge [116] created a digital signature over the root of the Merkle Hash Tree. Verification of a signature is a public key cryptography operation and consumes a large amount of energy. Our implementation of Elliptic Curve Digital Signature Algorithm (ECDSA) signatures over TelosB motes shows that verification of one signature needs 28.771 mJ of energy [83]. On the other hand, one AES-256 bit encryption costs .01 mJ of energy. In a traditional WSN, the digital signature is necessary because the entire network needs to be updated. In a sensor cloud, because only one cluster needs to be updated at a time, symmetric key cryptography can be used in place of public key cryptography. Instead of signing the root of the hash with the private key, the BS uses the session key (S_n) as a signature key. A signature packet (which includes the VHT root hash and the nonce n) is created and the signature is produced by encrypting (VHT root hash $||$ n). The nodes in the cluster can derive the session key (S_n) from the cluster key and the nonce and verify the root hash.

Index	CFL	BF	New Code	Key

Figure 7.4: Content disseminated by the base station.

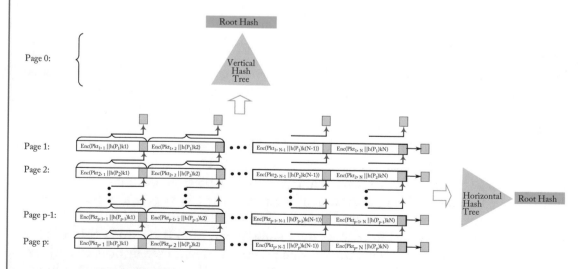

Figure 7.5: Vertical and horizontal hash trees.

We observe that code updating in a sensor cloud generally happens in a cluster where the sensors are typically one hop away from each other. The energy intensive tasks such as decryption of packets can therefore be performed in a distributed manner. Thus, instead of every sensor decrypting all the code, each sensor can decrypt a few packets which will result in conservation of a large amount of energy. This process, however, makes it necessary that nodes within a cluster are protected against malicious code injection from each other. To accomplish this, the base station creates another hash tree on the same dissemination contents, which we call the Horizontal Hash Tree (HHT). For this hash tree, each page of the code is hashed and the hashes of the pages $h(Page)$ are used as leaf nodes. The root hash of HHT is encrypted using the session key (S_n) and included in the signature packet. The code dissemination algorithm is given in Algorithm 7.11. The VHT and the HHT are illustrated in Figure 7.5. Just before beginning the code dissemination, BS broadcasts the next authentication key (A_i), which was used to create the HMAC in the pre-dissemination phase.

Algorithm 7.11 Code Dissemination

1: Populate BF by hashing the functions in CFL.
2: Create Index on $CFL||BF||NewCode||RK_i||K_i$.
3: Create session key $S_n = g_n(CK)$.
4: Encrypt packets of $Index||CFL||BF||NewCode||RK_i||K_i$ using S_i.
5: Hash the encrypted packets and create VHT with Page 0.
6: Hash the pages and create HHT.
7: Create the signature packet.
8: Broadcast Authentication Key A_i.
9: Disseminate Code.

7.7.4 ACTIVITY ON THE NODES

After receiving the next authentication key (A_i), the nodes verify this key by determining or not whether $h(A_i) = A_{i-1}$. The one way property of the hash chain ensures that any malicious node, which has obtained previous authentication keys, $A_{i-1}, A_{i-2}, \ldots$, etc. cannot predict the key A_i, with non-negligible probability. This implies that an adversary that makes any changes to the contents of the pre-dissemination packet, will be caught with a very high probability, which ensures the delivery of both the un-tampered $h(RK_i)$ and $h(CFL)$. The $h(RK_i)$ and $h(CFL)$ would be used to verify the re-encryption key and the CFL, which would be explained later in this subsection.

After the pre-dissemination phase is complete, the cluster for which the code dissemination was intended receives the encrypted contents. The contents are authenticated using the Vertical Hash Tree, in a manner similar to that used by Seluge [116]. The session key (S_n) is derived by using the PRF $g()$ on the cluster key CK, with nonce n received in the signature packet.

To enable distributed decryption and authentication of the contents, the cluster head in each cluster creates virtual ids ranging from 1 to N and gives each sensor one of the virtual ids, where N is the number of packets in a page. The sensors, instead of dealing with all the packets, only store the packets which are multiples of their virtual id. Thus, a node with virtual id 1, decrypts $pkt1$ in all the pages. Likewise, the node with virtual id 2, decrypts $pkt2$ in all the pages and so on. When the number of nodes in the cluster is less than N, nodes can be given additional virtual ids. Each node can decrypt and authenticate the packet in page i from the hash in the packet in page $i + 1$ and finally packets in page 1 can be authenticated from the VHT. After all of the packets have been received and authenticated, the nodes encrypt their decrypted packets again and broadcast for other nodes to receive. The encryption is done in large blocks, reducing the number of encryption and decryption operations a node must perform and thus conserving energy on nodes.

Once the nodes receive all the packets from the code dissemination, they first verify that the cluster members have not tampered with the code. This is done by hashing each page and verifying the root of the HHT. Since nodes are only allowed to decrypt a part of each page, any change in the code by a malicious node will always be identified. After the verification phase is complete, the nodes use the index to extract the CFL, the Bloom filter, the new code, the re-encryption key RK_i, and the encryption key K_i. The cluster head then broadcasts RK_i and CFL in clear.

When a node receives this broadcast packet, it checks if it has one or more of the requested functions in the CFL by comparing the FIDs. If the node finds that it has some of the requested functions, it verifies the validity of the CFL and RK_i by creating an HMAC over $h(CFL)$ and $h(RK_i)$ using the authentication key A_i. This HMAC is compared with the HMAC received in the pre-dissemination phase. If both of the HMACs match, CFL and the RK_i are valid. The requested functions are then re-encrypted with RK_i and sent back to the requesting nodes. The encryption key (K_i) received by the cluster nodes in the code dissemination is used to decrypt the received encrypted functions. The received functions are then verified using the Bloom Filter. The functions are hashed and the resulting positions in the filter are checked against the already existing entries in the Bloom Filter. If the hashes of a received function result in positions which are unset in the Bloom Filter, the function is rejected, otherwise it is accepted. Once all functions pass through the Bloom Filter, the nodes are ready to build the code image from its various parts. To build the code image, the nodes use the CFL to plug the common functions into their appropriate position in the code. The code image is stored and built in the flash memory. Once the build is complete, the bootloader can boot the node up using this image. The entire process is illustrated in Algorithm 7.12.

Algorithm 7.12 Image Build

1: Cluster head assigns virtual ids to sensors.
2: Nodes in the cluster store dissemination packets corresponding to their virtual ids.
3: Encrypted packets are verified using hashes and then decrypted.
4: Decrypted packets with hashes are combined in large blocks and securely broadcasted in the cluster.
5: All nodes receive all the transmitted code.
6: VHT and HHT are verified.
7: Base station broadcasts the CFL.
8: Nodes receive encrypted common functions from other nodes.
9: Functions are decrypted and verified through BF.
10: **if** functions are verified **then**
11:     Build code image.
12:     **if** all functions are verified **then**
13:         Reboot with new code.
14:     **end if**
15: **end if**

7.8   A DISCUSSION ON SECURITY

Our algorithm divides an application's code into two parts: new code and the common code. Different mechanisms are used for the security of both these categories, which are discussed in the following subsections.

7.8.1   CONFIDENTIALITY OF CODE

New Code

Confidentiality of the new code is provided by encrypting the dissemination content with the session key (S_n). This key is created uniquely for each session by using the cluster's key CK and a randomly generated nonce (n). This nonce is sent in the signature packet with the VHT root hash, and the signature over (VHT root hash$||n$). The signature guarantees the correct reception of both the VHT root hash and the nonce n.

Common Code

Confidentiality of the common code is provided by proxy re-encryption. The following discussion is based on the EC-BBS proxy re-encryption scheme, although a similar argument can also be made using SRE [119]. In the pre-deployment phase, the base station encrypts randomly chosen functions with the encryption key Q_0 using EC-BBS and stores them on the nodes. Neither this key Q_0 nor the corresponding secret key K_0 are revealed to any node. This provides the confidentiality of the common functions from the nodes on which the functions are stored. Prior to disseminating any application code, the base station first broadcasts $HMAC_{A_i}(h(RK_i)||h(CFL))$ to all nodes. The first packet in the code dissemination is the packet that contains the authentication key A_i. The validity of the authentication key can be verified by determining whether or not $h(A_i) = A_{i-1}$. After code dissemination, when node j requests a function from node k, j needs to send the CFL and the re-encryption key RK_i to node k. Node k checks the validity of the request by generating an HMAC on the received CFL and RK_i, and comparing it with the HMAC received in the pre-dissemination phase. It then re-encrypts the requested function using the received RK_i. This process allows node j to receive re-encrypted functions without revealing them to node k or any other unauthorized node.

7.8.2   INTEGRITY OF CODE

New Code

For integrity of the new code we use two hash trees: the VHT and the HHT. The hashes of packets in Page 1 are used as leaf nodes and the resulting tree is taken as the VHT. The VHT is used in the same manner as that described by Seluge [116]. For the HHT, each page is hashed and the hashes of pages are taken as the leaf nodes and a tree is built. When the nodes in a cluster perform distributed decryption of the disseminated code, corrupt nodes can inject malicious code into the code image. The HHT, however, makes sure that such a code injection will always be detected.

Common Code

When node j requests a function from node k, a corrupt k can return corrupt or malicious code back to j. Node j verifies all the received functions against the Bloom Filter it received during code dissemination to thwart such attacks.

7.9   SUMMARY

Sensor clouds are an emerging paradigm for sensor networks. These are very dynamic in nature with nodes being constantly provisioned and de-provisioned for users. In such a scenario, an efficient code dissemination algorithm that is also secure becomes necessary. In this book, we have presented a novel code dissemination algorithm that is both efficient and secure. Our code dissemination algorithm considers the similarities that exist between codes across applications. The basic idea, therefore, is to communicate only the new code to the sensors while the common code can be picked up from the sensors in the network. This process reduces the amount of code that needs to be communicated. A reduced amount of code results in energy efficient code dissemination. Our security framework is based around proxy re-encryption, hash trees, and Bloom Filters, which combine to provide confidentiality and integrity to the code dissemination algorithm.

Bibliography

[1] D. Boeringer and D. Werner, Particle swarm optimization vs. genetic algorithms for phased array synthesis, *IEEE Transactions on Antennas and Propagation*, vol. 52, no. 3, pp. 771–779, 2004. DOI: 10.1109/tap.2004.825102.

[2] J. Carroll, E. Briscoe, and A. Sanfilippo, Supertagging: An approach to almost parsing, *Computational Linguistics*, 25(2):237–267.

[3] S. Ho, S. Yang, G. Ni, E. W. C. Lo, and H. C. Wong, A particle swarm optimization-based method for multiobjective design optimizations, *IEEE Transactions on Magnetics*, vol. 41, no. 5, pp. 1756–1759, 2005. DOI: 10.1109/tmag.2005.846033.

[4] H. Kopka, and P. W. Daly, *Guide to LaTeX*, 4th ed., Addison Wesley.

[5] F. S. Levin, *An Introduction to Quantum Theory*, Cambridge, Cambridge University Press, 2002. DOI: 10.1017/cbo9781139164177.

[6] K. Tarvainen, *Einführung in die Dependenzgrammatik*, Niemeyer. DOI: 10.1515/9783110920673.

[7] J. Lu, H. Okada, T. Itoh, R. Maeda, and T. Harada, Towards the world smallest wireless sensor nodes with low power consumption for Green Sensor Networks, *IEEE SENSORS*, pp. 1–4, 2013 DOI: 10.1109/icsens.2013.6688357. 1

[8] H. Gao, Fully integrated ultra-low power mm-wave wireless sensor design methods, Doctoral thesis, 2015. 1

[9] A. Sorniotti, L. Gomez, K. Wrona, and L. Odorico, Secure and trusted in-network data processing in wireless sensor networks: A survey, *Journal of Information Assurance and Security*, vol. 2, no. 3, pp. 189–199, 2007. 12

[10] R. Pereira, J. Trindade, F. Goncalves, L. Suresh, D. Barbosa, and T. Vazao, A wireless sensor network for monitoring volcano-seismic signals, *Natural Hazards and Earth System Sciences*, vol. 14, no. 12, pp. 3123–3142, 20014. DOI: 10.5194/nhess-14-3123-2014. 2

[11] S. Bhatti, J. Carlson, H. Dai, J. Deng, J. Rose, A. Sheth, B. Shucker, C. Gruenwald, A. Torgerson, and R. Han, MANTIS OS: An embedded multithreaded operating system for wireless micro sensor platforms, *Mobile Networks and Applications*, vol. 10, no. 4, pp. 563–579, 2005. DOI: 10.1007/s11036-005-1567-8. 7

[12] N. Brouwers, K. Langendoen, and P. Corke, Darjeeling, a feature-rich VM for the resource poor, *Proc. of the 7th ACM Conference on Embedded Networked Sensor Systems*, pp. 169–182, 2009. DOI: 10.1145/1644038.1644056. 7

[13] A. P. Jayasumana, Q. Han, and T. H. Illangasekare, Virtual sensor networks—a resource efficient approach for concurrent applications, *4th International Conference on Information Technology*, pp. 111–115, 2007. DOI: 10.1109/itng.2007.206. 4

[14] L. Gheorghe, D. Tudose, M. Wehner, and S. Zeisberg, Securing virtual networks for multi-owner wireless sensor networks, *4th International Conference on Intelligent Networking and Collaborative Systems*, pp. 630–635, 2012. DOI: 10.1109/incos.2012.91. 4

[15] A. Kapadia, S. Myers, X. Wang, and G. Fox, Secure cloud computing with brokered trusted sensor networks, *International Symposium on Collaborative Technologies and Systems*, pp. 581–592, 2010. DOI: 10.1109/cts.2010.5478459. 4

[16] S. Kamburugamuve, L. Christiansen, and G. Fox, A framework for real time processing of sensor data in the cloud, *Journal of Sensors*, 2015. DOI: 10.1155/2015/468047. 4

[17] J. Koshy and R. Pandey, VM*: Synthesizing scalable runtime environments for sensor networks, *Proc. of the 3rd International Conference on Embedded Networked Sensor Systems*, pp. 243–254, ACM, 2005. DOI: 10.1145/1098918.1098945. 7

[18] P. Lewis and D. Culler, Maté: A tiny virtual machine for sensor networks, *Proc. of the 10th International Conference on Architectural Support for Programming Languages and Operating Systems*, pp. 85–95, ACM, 2002. DOI: 10.1145/605397.605407. 6

[19] S. Madria, V. Kumar, and R. Dalvi Sensor cloud: A cloud of virtual sensors, *IEEE Software*, vol. 31, no. 2, pp. 70–77, 2014. DOI: 10.1109/ms.2013.141. 1, 4, 7, 45

[20] S. Michiels, W. Horré, W. Joosen, and P. Verbaeten, DAViM: A dynamically adaptable virtual machine for sensor networks, *Proc. of the International Workshop on Middleware for Sensor Networks*, pp. 7–12, 2006. DOI: 10.1145/1176866.1176868. 7

[21] J. Polastre, R. Szewczyk, and D. Culler, Telos: Enabling ultra-low power wireless research, in *Proc. of the 4th International Symposium on Information Processing in Sensor Networks, (IPSN)*, Piscataway, NJ, IEEE Press, 2005. http://dl.acm.org/citation.cfm?id=1147685.1147744 DOI: 10.1109/ipsn.2005.1440950. 35

[22] S. S. K. Dash and P. Pattnaik, A survey on applications of wireless sensor network using cloud computing, *International Journal of Computer Science and Emerging Technologies*, vol. 1, no. 4, pp. 50–55, December 2010.

[23] P. Levis, S. Madden, J. Polastre, R. Szewczyk, K. Whitehouse, A. Woo, D. Gay, J. Hill, M. Welsh, E. Brewer, and D. Culler, Tinyos: An operating system for sensor networks, in *Ambient Intelligence*, W. Weber, J. Rabaey, and E. Aarts, Eds., pp. 115–148, Springer Berlin Heidelberg, 2005. DOI: 10.1007/3-540-27139-2_7. 38

[24] L. M. Surhone, M. T. Tennoe, and S. F. Henssonow, *Node.Js*, Mauritius, Betascript Publishing, 2010. 38

[25] V. Kumar and S. K. Madria, Efficient and secure code dissemination in sensor clouds, in *IEEE 15th International Conference on Mobile Data Management, (MDM)*, vol. 1, pp. 103–112, Brisbane, Australia, July 14–18, 2014. DOI: 10.1109/mdm.2014.19.

[26] V. Kumar and S. Madria, Pip: Privacy and integrity preserving data aggregation in wireless sensor networks, in *IEEE 32nd Symposium on Reliable Distributed Systems, (SRDS)*, pp. 10–19, Braga, Portugal, October 1–3, 2013. DOI: 10.1109/srds.2013.10.

[27] R. Dalvi and S. K. Madria, Energy efficient scheduling of fine-granularity tasks in a sensor cloud, in *Database Systems for Advanced Applications—20th International Conference, (DASFAA), Proceedings, Part II*, pp. 498–513, Hanoi, Vietnam, April 20–23, 2015. DOI: 10.1007/978-3-319-18123-3_30. 42

[28] G. Kortuem, F. Kawsar, V. Sundramoorthy, and D. Fitton, Smart objects as building blocks for the Internet of Things, in *IEEE Internet Computing*, 14(1), 44–51, January 2010. DOI: 10.1109/mic.2009.143. 27

[29] M. M. Hassan, B. Song, and E.-N. Huh, A framework of sensor-cloud integration opportunities and challenges, in *Proc. of the 3rd International Conference on Ubiquitous Information Management and Communication, (ICUIMC)*, pages 618–626, Suwon, Korea, 2009. DOI: 10.1145/1516241.1516350. 27

[30] S. Kabadayi, A. Pridgen, and C. Julien, Virtual sensors: Abstracting data from physical sensors, in *Proc. of the International Symposium on on World of Wireless, Mobile and Multimedia Networks, (WOWMOM)*, pages 587–592, Washington, DC, 2006. DOI: 10.1109/wowmom.2006.115. 27

[31] N. K. Kapoor, S. Majumdar, and B. Nandy, Scheduling on wireless sensor networks hosting multiple applications, in *IEEE International Conference on Communications, (ICC)*, pages 1–6, June 2011. DOI: 10.1109/icc.2011.5963195. 27

[32] K. Aberer, M. Hauswirth, and A. Salehi, A Middleware for fast and flexible sensor network deployment, in *Proc. of the 32nd International Conference on Very Large Data Bases, (VLDB)*, pages 1199–1202, Seoul, Korea, 2006. 27

[33] N. Poolsappasit, V. Kumar, S. Madria, and S. Chellappan, Challenges in secure sensor-cloud computing, in *Proc. of the 8th VLDB International Conference on Secure Data Management, (SDM)*, pp. 70–84, Berlin, Heidelberg, Springer-Verlag, 2011, DOI: 10.1007/978-3-642-23556-6_5. 27, 45

[34] A. Kapadia, S. Myers, X. Wang, and G. Fox, Toward securing sensor clouds, in *Collaboration Technologies and Systems (CTS), International Conference on*, pp. 280–289, 2011. DOI: 10.1109/cts.2011.5928699. 45

[35] C. W. K. Pongaliur and L. Xiao, Maintaining functional module integrity in sensor networks, *IEEE International Conference on Mobile Ad hoc and Sensor Systems Conference*, vol. 0, p. 806, 2005. DOI: 10.1109/mahss.2005.1542874. 46, 48

[36] E.-H. Ngai, J. Liu, and M. Lyu, On the intruder detection for sinkhole attack in wireless sensor networks, in *Communications, (ICC). IEEE International Conference on*, vol. 8, pp. 3383–3389, 2006. DOI: 10.1109/icc.2006.255595. 46

[37] J. Newsome, E. Shi, D. Song, and A. Perrig, The sybil attack in sensor networks: Analysis defenses, in *Information Processing in Sensor Networks, (IPSN). 3rd International Symposium on*, pp. 259–268, 2004. DOI: 10.1145/984622.984660. 18, 46

[38] J. P. Walters, Z. Liang, W. Shi, and V. Chaudhary, Wireless sensor network security: A survey, in book chapter of security, in *Distributed, Grid, and Pervasive Computing*, Yang Xiao Ed., CRC Press, pp. 0–849, 2007. 18, 48, 50

[39] Y Wang, G. Attebury, and B. Ramamurthy, A survey of security issues in wireless sensor networks, in *IEEE Communications Surveys Tutorials*, vol. 8, pp. 2–23, 2006. DOI: 10.1109/comst.2006.315852. 50

[40] I. Ray and N. Poolsapassit, Using attack trees to identify malicious attacks from authorized insiders, in *Proc. of the 10th European Conference on Research in Computer Security, (ESORICS)*, pp. 231–246, Berlin, Heidelberg, Springer-Verlag, 2005. DOI: 10.1007/11555827_14. 46

[41] J. Dawkins, C. Campbell, and J. Hale, Modeling network attacks: Extending the attack tree paradigm, in *Workshop on Statistical and Machine Learning Techniques in Computer Intrusion Detection*, pp. 75–86, 2002. 46

[42] N. Poolsappasit, R. Dewri, and I. Ray, Dynamic security risk management using Bayesian attack graphs, *IEEE Transactions on Dependable and Secure Computing*, vol. 9, no. 1, pp. 61–74, January 2012. DOI: 10.1109/tdsc.2011.34. 46

[43] A. Terje, Trends in quantitative risk assessments, *International Journal of Performability Engineering*, vol. 5, no. 5, pp. 447–461, October 2009.

[44] The national vulnerability database (NVD), 2005. https://nvd.nist.gov/ 19, 47

[45] M. Frigault and L. Wang, Measuring network security using Bayesian network-based attack graphs, in *Proc. of the 32nd Annual IEEE International Computer Software and Applications Conference, (COMPSAC)*, pp. 698–703, Washington, DC, Computer Society, 2008. DOI: 10.1109/compsac.2008.88. 19, 47

[46] S. Houmb and V. Nunes Leal Franqueira, Estimating toe risk level using CVSS, in *Proc. of the 4th International Conference on Availability, Reliability and Security (ARES The International Dependability Conference)*, pp. 718–725, IEEE Conference Proceedings, Los Alamitos, IEEE Computer Society Press, March 2009. DOI: 10.1109/ares.2009.151. 19, 48, 55

[47] Nessus vulnerability scanner v6.0, 2014. https://www.tenable.com/products/nessus-vulnerability-scanner 20, 46

[48] C. Mockel and A. E. Abdallah, Threat modeling approaches and tools for securing architectural designs of an e-banking application, in *Information Assurance and Security (IAS), 6th International Conference on*, pp. 149–154, August 2010. DOI: 10.1109/isias.2010.5604049. 20

[49] O. Sheyner and J. Wing, Tools for generating and analyzing attack graphs, in *Proc. of Formal Methods for Components and Objects, Lecture Notes in Computer Science*, pp. 344–371, 2004. DOI: 10.1007/978-3-540-30101-1_17. 19, 47

[50] C. Phillips, A graph-based system for network-vulnerability analysis, in *Proc. of the Workshop on New Security Paradigms*, pp. 71–79, ACM Press, 1998. DOI: 10.1145/310889.310919. 19, 48

[51] S. Barnum, Attack patterns: Knowing your enemy in order to defeat them, *BlackHat DC, cigital*, www.cigital.com, 2007. 49, 50

[52] G. Padmavathi and D. Shanmugapriya, A survey of attacks, security mechanisms and challenges in wireless sensor networks, *CoRR*, vol. abs/0909.0576, 2009. 18, 50

[53] Y. Wang, G. Attebury, and B. Ramamurthy, A survey of security issues in wireless sensor networks, *IEEE Communications Surveys Tutorials*, vol. 8, pp. 2–23, 2006. DOI: 10.1109/comst.2006.315852. 18

[54] P. Mell, K. Scarfone, and S. Romanosky, *A Complete Guide to the Common Vulnerability Scoring System Version 2.0*, 1st ed., NIST and Carnegie Mellon University, June 2007. http://www.first.org/cvss/v2/guide 21, 55

[55] Seamonster—security modeling software, 2013. www.sourceforge.net/projects/seamonster

[56] E. Shi and A. Perrig, Designing secure sensor networks, *Wireless Communications, IEEE*, vol. 11, no. 6, pp. 38–43, 2004. DOI: 10.1109/mwc.2004.1368895. 18

[57] A. Wood and J. Stankovic, Denial of service in sensor networks, *Computer*, vol. 35, no. 10, pp. 54–62, 2002. DOI: 10.1109/mc.2002.1039518. 18

[58] W. Xu, K. Ma, W. Trappe, and Y. Zhang, Jamming sensor networks: Attack and defense strategies, *IEEE Network*, vol. 20, no. 3, pp. 41–47, 2006. DOI: 10.1109/mnet.2006.1637931. 18

[59] C. Karlof and D. Wagner, Secure routing in wireless sensor networks: Attacks and countermeasures, in *1st IEEE International Workshop on Sensor Network Protocols and Applications*, pp. 113–127, 2002. DOI: 10.1109/snpa.2003.1203362. 18

[60] B. Kannhavong, H. Nakayama, Y. Nemoto, N. Kato, and A. Jamalipour, A survey of routing attacks in mobile ad hoc networks, *Wireless Communications, IEEE*, vol. 14, no. 5, pp. 85–91, 2007. DOI: 10.1109/mwc.2007.4396947. 18

[61] S. Mauw and M. Oostdijk, Foundations of attack trees, in *ICISC*, pp. 186–198, 2005. DOI: 10.1007/11734727_17. 19

[62] J. Lee, H. Lee, and H. P. In, Scalable attack graph for risk assessment, in *Proc. of the 23rd International Conference on Information Networking, (ICOIN)*, pp. 78–82, IEEE Press, Piscataway, NJ, 2009. 19

[63] L. Gallon and J. J. Bascou, Using CVSS in attack graphs, in *Proc. of the 6th International Conference on Availability, Reliability and Security, (ARES)*, pp. 59–66, IEEE Computer Society, Washington, DC, 2011. DOI: 10.1109/ares.2011.18. 19

[64] R. Dantu, K. Loper, and P. Kolan, Risk management using behavior based attack graphs, in *Proc. of the International Conference on Information Technology: Coding and Computing (ITCC) Volume 2*, pp. 445–, IEEE Computer Society, Washington, DC, 2004. DOI: 10.1109/itcc.2004.1286496. 19

[65] Y. Liu and H. Man, Network vulnerability assessment using Bayesian networks, vol. 5812, pp. 61–71, 2005. DOI: 10.1117/12.604240. 19

[66] C. Gentry, Fully homomorphic encryption using ideal lattices, in *Proc. of the 41st Annual ACM Symposium on Theory of Computing, (STOC)*, pp. 169–178, 2009. DOI: 10.1145/1536414.1536440. 22

[67] S. Blake-Wilson, N. Bolyard, V. Gupta, C. Hawk, and B. Moeller, Elliptic curve cryptography (ECC) cipher suites for transport layer security (TLS), RFC 4492, Networks Working Group, 2006. DOI: 10.17487/rfc4492. 23

[68] J. Albath and S. Madria, Secure hierarchical data aggregation in wireless sensor networks, in *WCNC*, pp. 2420–2425, IEEE, 2009. DOI: 10.1109/wcnc.2009.4917960. 70

[69] L. Hu and D. Evans, Secure aggregation for wireless networks, in *Workshop on Security and Assurance in Ad Hoc Networks*, IEEE Computer Society, 2003. DOI: 10.1109/saintw.2003.1210191. 70

[70] H. Chan, A. Perrig, and D. X. Song, Secure hierarchical in-network aggregation in sensor networks, in *Computer and Communications Security, (CCS)*, pp. 278–287, 2006. DOI: 10.1145/1180405.1180440. 70

[71] A. Mahimkar and T. S. Rappaport, SecureDAV: A secure data aggregation and verification protocol for sensor networks, in *Proc. of the IEEE Global Telecommunications Conference*, pp. 2175–2179, 2004. DOI: 10.1109/glocom.2004.1378395. 70

[72] Y. Yang, X. Wang, S. Zhu, and G. Cao, SDAP: A secure hop-by-hop data aggregation protocol for sensor networks, in *Proc. of the 7th ACM International Symposium on Mobile Ad Hoc Networking and Computing, (MobiHoc)*, pp. 356–367, 2006. DOI: 10.1145/1132905.1132944. 70

[73] J. Bahi, C. Guyeux, and A. Makhoul, Efficient and robust secure aggregation of encrypted data in sensor networks, in *4th International Conference on Sensor Technologies and Applications, (SENSORCOMM)*, pp. 472–477, July 2010. DOI: 10.1109/sensorcomm.2010.76. 70, 71

[74] C. Castelluccia, E. Mykletun, and G. Tsudik, Efficient aggregation of encrypted data in wireless sensor networks, in *Proc. of the 2nd Annual International Conference on Mobile and Ubiquitous Systems: Networking and Services, (Mobiquitous)*, pp. 109–117, IEEE Computer Society, 2005. DOI: 10.1109/mobiquitous.2005.25. 71

[75] H.-M. Sun, Y.-C. Hsiao, Y.-H. Lin, and C.-M. Chen, An efficient and verifiable concealed data aggregation scheme in wireless sensor networks, in *Proc. of the International Conference on Embedded Software and Systems*, pp. 19–26, IEEE Computer Society, 2008. DOI: 10.1109/icess.2008.9. 70, 71

[76] M. Groat, W. He, and S. Forrest, KIPDA: k-indistinguishable privacy preserving data aggregation in wireless sensor networks, in *INFOCOM*, pp. 2024–2032, 2011. DOI: 10.1109/infcom.2011.5935010.

[77] V. Kumar, and S. Madria, Secure data aggregation in wireless sensor networks, in *Wireless Sensor Network Technologies for the Information Explosion Era*, Springer, 2010. DOI: 10.1007/978-3-642-13965-9_3.

[78] P. L. Steffen Peter and K. Piotrowski, On concealed data aggregation for wireless sensor networks, in *Proc. of the 4th IEEE Consumer Communications and Networking Conference, (CCNC)*, January 2007.

[79] A. Liu and P. Ning, Tinyecc: A configurable library for elliptic curve cryptography in wireless sensor networks, in *Proc. of the 7th International Conference on Information Processing in Sensor Networks, (IPSN)*, pp. 245–256, IEEE Computer Society, 2008. DOI: 10.1109/ipsn.2008.47.

[80] C. Karlof, N. Sastry, and D. Wagner, Tinysec: A link layer security architecture for wireless sensor networks, in *Proc. of the 2nd International Conference on Embedded Networked Sensor Systems, (SenSys)*, pp. 162–175, ACM, 2004, DOI: 10.1145/1031495.1031515.

[81] G. de Meulenaer, F. Gosset, F. Standaert, and O. Pereira On the energy cost of communication and cryptography in wireless sensor networks, in *Proc. of the IEEE International Conference on Wireless and Mobile Computing, Networking and Communications, (WiMob)*, pp. 580–585, IEEE, 2008. DOI: 10.1109/wimob.2008.16. 69

[82] W. He, X. Liu, H. V. Nguyen, K. Nahrstedt, and T. Abdelzaher, Pda: Privacy-preserving data aggregation for information collection, *ACM Transactions on Sensor Networks*, vol. 8, no. 1, pp. 6:1–6:22, August 2011. DOI: 10.1145/1993042.1993048. 70, 71

[83] V. Kumar and S. K. Madria, Secure hierarchical data aggregation in wireless sensor networks: Performance evaluation and analysis, *Mobile Data Management, IEEE International Conference on*, vol. 0, pp. 196–201, 2012. DOI: 10.1109/mdm.2012.49. 108

[84] J. Girao, D. Westhoff, and M. Schneider, Cda: Concealed data aggregation for reverse multicast traffic in wireless sensor networks, in *Communications, IEEE International Conference on*, vol. 5, pp. 3044–3049, May 2005. DOI: 10.1109/icc.2005.1494953.

[85] T. Feng, C. Wang, W. Zhang, and L. Ruan, Confidentiality protection for distributed sensor data aggregation, in *INFOCOM. The 27th Conference on Computer Communications. IEEE*, pp. 56–60, 2008. DOI: 10.1109/infocom.2007.20. 70, 71

[86] L. Zhang, H. Zhang, M. Conti, R. Pietro, S. Jajodia, and L. Mancini, Preserving privacy against external and internal threats in WSN data aggregation, *Telecommunication Systems*, pp. 1–14, 2011. DOI: 10.1007/s11235-011-9539-8. 70, 71

[87] M. Conti, L. Zhang, S. Roy, R. Di Pietro, S. Jajodia, and L. V. Mancini, Privacy-preserving robust data aggregation in wireless sensor networks, *Security and Communication Networks*, vol. 2, no. 2, pp. 195–213, 2009. DOI: 10.1002/sec.95.

[88] X. Lin, R. Lu, and X. S. Shen, MDPA: Multidimensional privacy-preserving aggregation scheme for wireless sensor networks, *Wireless Communications Mobile Computing*, vol. 10, no. 6, pp. 843–856, June 2010. DOI: 10.1002/wcm.796.

[89] C. Wang, G. Wang, W. Zhang, and T. Feng, Reconciling privacy preservation and intrusion detection in sensory data aggregation, in *INFOCOM, Proceedings IEEE*, 2011. DOI: 10.1109/infcom.2011.5935177.

[90] W. He, H. Nguyen, X. Liu, K. Nahrstedt, and T. Abdelzaher, IPDA: An integrity-protecting private data aggregation scheme for wireless sensor networks, in *MILCOM. IEEE*, pp. 1–7, November 2008. DOI: 10.1109/milcom.2008.4753645.

[91] W. He, X. Liu, H. Nguyen, and K. Nahrstedt, A cluster-based protocol to enforce integrity and preserve privacy in data aggregation, in *Distributed Computing Systems Workshops, (ICDCS Workshops), 29th IEEE International Conference on*, pp. 14–19, June 2009. DOI: 10.1109/icdcsw.2009.18.

[92] A. Shamir, How to share a secret, *Communications of the ACM*, vol. 22, no. 11, pp. 612–613, November 1979. DOI: 10.1145/359168.359176. 25

[93] A. Parakh and S. Kak, Recursive secret sharing for distributed storage and information hiding, *ANTS*, pp. 88–90, IEEE Press, Piscataway, NJ, 2009. DOI: 10.1109/ants.2009.5409868. 75

[94] C. Castelluccia, A. C.-F. Chan, E. Mykletun, and G. Tsudik, Efficient and provably secure aggregation of encrypted data in wireless sensor networks, *ACM Transactions on Sensor Networks*, vol. 5, no. 3, pp. 20:1–20:36, June 2009. DOI: 10.1145/1525856.1525858.

[95] M. Yuriyama and T. Kushida, Sensor-cloud infrastructure-physical sensor management with virtualized sensors on cloud computing, in *Network-Based Information Systems (NBiS), 13th International Conference on*, pp. 1–8, IEEE, 2010. DOI: 10.1109/nbis.2010.32. 97

[96] N. Poolsappasit, V. Kumar, S. Madria, and S. Chellappan, Challenges in secure sensor-cloud computing, *Secure Data Management*, pp. 70–84, 2011. DOI: 10.1007/978-3-642-23556-6_5. 81, 97

[97] S. Yu, K. Ren, and W. Lou, FDAC: Toward fine-grained distributed data access control in wireless sensor networks, *IEEE Transactions on Parallel and Distributed Systems*, vol. 22, no. 4, pp. 673–686, 2011. DOI: 10.1109/tpds.2010.130. 81, 82, 83, 84, 91, 92

[98] S. Ruj, A. Nayak, and I. Stojmenovic, Distributed fine-grained access control in wireless sensor networks, in *Parallel Distributed Processing Symposium (IPDPS), IEEE International*, pp. 352–362, 2011. DOI: 10.1109/ipdps.2011.42. 81, 82, 83, 84

[99] D. Liu, Efficient and distributed access control for sensor networks, in *Distributed Computing in Sensor Systems*, Lecture Notes in Computer Science, J. Aspnes, C. Scheideler, A. Arora, and S. Madden, Eds., vol. 4549, pp. 21–35, Springer Berlin Heidelberg, 2007. DOI: 10.1007/978-3-540-73090-3_2. 82

[100] J. Maerien, S. Michiels, C. Huygens, D. Hughes, and W. Joosen, Access control in multi-party wireless sensor networks, in *Wireless Sensor Networks*, Lecture Notes in Computer Science, P. Demeester, I. Moerman, and A. Terzis, Eds., vol. 7772, pp. 34–49, Springer Berlin Heidelberg, 2013. DOI: 10.1007/978-3-642-36672-7_3. 82

[101] H. Wang and Q. Li, Achieving distributed user access control in sensor networks, *Ad Hoc Networks*, vol. 10, no. 3, pp. 272–283, 2012. DOI: 10.1016/j.adhoc.2011.01.011. 82

[102] G. Bianchi, A. T. Capossele, C. Petrioli, and D. Spenza, Agree: Exploiting energy harvesting to support data-centric access control in {WSNs}, *Ad Hoc Networks*, vol. 11, no. 8, pp. 2625–2636, 2013. DOI: 10.1016/j.adhoc.2013.03.013. 82

[103] J. Bethencourt, A. Sahai, and B. Waters, Ciphertext-policy attribute-based encryption, in *Security and Privacy, (SP). IEEE Symposium on*, pp. 321–334, 2007. DOI: 10.1109/sp.2007.11. 83, 92

[104] A. Sahai and B. Waters, Fuzzy identity-based encryption, in *Advances in Cryptology, (EUROCRYPT)*, Lecture Notes in Computer Science, vol. 3494, pp. 457–473, Springer Berlin Heidelberg, 2005. DOI: 10.1007/11426639_27. 83

[105] V. Goyal, O. Pandey, A. Sahai, and B. Waters, Attribute-based encryption for fine-grained access control of encrypted data, in *Proc. of the 13th ACM Conference on Computer and Communications Security, (CCS)*, pp. 89–98, 2006. DOI: 10.1145/1180405.1180418. 23, 83, 84, 90, 95

[106] P. Paillier, Public-key cryptosystems based on composite degree residuosity classes, in *Advances in Cryptology, (EUROCRYPT)*, Lecture Notes in Computer Science, vol. 1592, pp. 223–238, Springer Berlin Heidelberg, 1999. DOI: 10.1007/3-540-48910-x_16. 22, 90

[107] V. Kumar and S. Madria, PIP: Privacy and integrity preserving data aggregation in wireless sensor networks, in *Reliable Distributed Systems (SRDS), IEEE 32nd International Symposium on*, pp. 10–19, 2013. DOI: 10.1109/srds.2013.10. 79

[108] X. Xiong, D. S. Wong, and X. Deng, Tinypairing: A fast and lightweight pairing-based cryptographic library for wireless sensor networks, in *Wireless Communications and Networking Conference (WCNC), IEEE*, pp. 1–6, 2010. DOI: 10.1109/wcnc.2010.5506580.

[109] N. Bin Shafi, K. Ali, and H. Hassanein, No-reboot and zero-flash over-the-air programming for wireless sensor networks, in *Sensor, Mesh and Ad Hoc Communications and Networks (SECON), 9th Annual IEEE Communications Society Conference on*, pp. 371–379, June 2012. DOI: 10.1109/secon.2012.6275799. 97, 99, 100, 103, 104

[110] W. Dong, Y. Liu, C. Chen, J. Bu, C. Huang, and Z. Zhao, R2: Incremental reprogramming using relocatable code in networked embedded systems, *IEEE Transactions on Computers*, vol. 99, PrePrints, 2012. DOI: 10.1109/tc.2012.161.

[111] J. Jeong and D. Culler, Incremental network programming for wireless sensors, in *Sensor and Ad Hoc Communications and Networks, (SECON). 1st Annual IEEE Communications Society Conference on*, pp. 25–33, October 2004. DOI: 10.1109/sahcn.2004.1381899. 97, 99

[112] J. Koshy and R. Pandey, Remote incremental linking for energy-efficient reprogramming of sensor networks, in *Wireless Sensor Networks. Proc. of the 2nd European Workshop on*, pp. 354–365, 2005. DOI: 10.1109/ewsn.2005.1462027. 99

[113] N. Reijers and K. Langendoen, Efficient code distribution in wireless sensor networks, in *Proc. of the 2nd ACM International Conference on Wireless Sensor Networks and Applications, (WSNA)*, pp. 60–67, New York, 2003. DOI: 10.1145/941350.941359. 97, 98, 99, 100

[114] J. Hui and D. Culler, The dynamic behavior of a data dissemination protocol for network programming at scale, in *Proc. of the 2nd International Conference on Embedded Networked Sensor Systems*, pp. 81–94, ACM, 2004. DOI: 10.1145/1031495.1031506. 98, 99

[115] W. Li, Y. Du, Y. Zhang, B. R. Childers, P. Zhou, and J. Yang, Adaptive buffer management for efficient code dissemination in multi-application wireless sensor networks, in *IEEE/IPIP International Conference on Embedded and Ubiquitous Computing*, pp. 295–301, 2008. DOI: 10.1109/euc.2008.160. 99

[116] S. Hyun, P. Ning, A. Liu, and W. Du, Seluge: Secure and dos-resistant code dissemination in wireless sensor networks, in *Proc. of the 7th International Conference on Information Processing in Sensor Networks*, pp. 445–456, IEEE Computer Society, 2008. DOI: 10.1109/ipsn.2008.12. 99, 100, 107, 108, 109, 112

[117] P. K. Dutta, J. W. Hui, D. C. Chu, and D. E. Culler, Securing the deluge network programming system, in *Proc. of the 5th International Conference on Information Processing in Sensor Networks, (IPSN)*, pp. 326–333, New York, ACM, 2006. DOI: 10.1109/ipsn.2006.243821. 100

[118] H. Tan, D. Ostry, J. Zic, and S. Jha, A confidential and dos-resistant multi-hop code dissemination protocol for wireless sensor networks, *Computer and Security*, vol. 32, no. 0, pp. 36–55, 2013. DOI: 10.1145/1514274.1514308. 100

[119] A. Syalim, T. Nishide, and K. Sakurai, Realizing proxy re-encryption in the symmetric world, in *Informatics Engineering and Information Science*, Communications in Computer

and Information Science, A. Abd Manaf, A. Zeki, M. Zamani, S. Chuprat, and E. El-Qawasmeh, Eds., vol. 251, pp. 259–274, 2011. DOI: 10.1007/978-3-642-25327-0_23. 24, 100, 102, 111

[120] M. Blaze, G. Bleumer, and M. Strauss, Divertible protocols and atomic proxy cryptography, in *EUROCRYPT*, pp. 127–144, 1998. DOI: 10.1007/bfb0054122. 24, 100, 102

[121] J. Albath, M. Thakur, and S. Madria, Energy constrained dominating set for clustering in wireless sensor networks, *IEEE 27th International Conference on Advanced Information Networking and Applications (AINA)*, pp. 812–819, 2010. DOI: 10.1109/aina.2010.14. 100

[122] B. H. Bloom, Space/time trade-offs in hash coding with allowable errors, *Communications of the ACM*, vol. 13, no. 7, pp. 422–426, July 1970. DOI: 10.1145/362686.362692. 25

[123] T. Eisenbarth, R. Koschke, and D. Simon, Aiding program comprehension by static and dynamic feature analysis, in *Software Maintenance, Proceedings. IEEE International Conference on*, pp. 602–611, 2001. DOI: 10.1109/icsm.2001.972777. 104

[124] Y. Mao, F. Wang, L. Qiu, S. Lam, and J. Smith, S4: Small state and small stretch compact routing protocol for large static wireless networks, *Networking, IEEE/ACM Transactions on*, vol. 18, no. 3, pp. 761–774, 2010. DOI: 10.1109/tnet.2010.2046645.

[125] M. Stanley and J. Lee, Sensor analysis for the Internet of Things, *Synthesis Lectures*, Morgan & Claypool Publishers, March 2018. 2

[126] I. F. Akyildiz, W. Su, Y. Sankarasubramaniam, and E. Cayirci, Wireless sensor networks: A survey, *Computer Networks*, vol. 38, no. 4, pp. 393–422, 2002. 2

[127] M. Luk, G. Mezzour, A. Perrig, and V. Gligor, MiniSec: A secure sensor network communication architecture, *Proc. of the 6th International Symposium on Information Processing in Sensor Networks*, pp. 479–488, 2007. 2

Authors' Biographies

VIMAL KUMAR

Vimal Kumar is a lecturer with the Computer Science Department of the Faculty of Computing and Mathematical Sciences at the University of Waikato in Hamilton, New Zealand. He received his Ph.D. from Missouri University of Science and Technology, USA, in 2013 and was previously Assistant Professor at the University of West Florida and software engineer for Computer Sciences Corporation. His research interests are broadly in the areas of wireless sensors, cloud computing, and security. He is particularly interested in the development of secure wireless sensor networks, sensor clouds, and the "Internet of Things" and is also looking into the security of wearable devices and other issues in secure cloud computing.

AMARTYA SEN

Amartya Sen is an Assistant Professor in the Department of Computer Science and Engineering at Oakland University, Rochester, Michigan. He received his Ph.D. in Computer Science from Missouri University of Science and Technology, USA in May 2018. He has published in peer-reviewed journals such as *IEEE Transaction on Services Computing* and *IEEE Cloud Computing*. His research interests are in the broad areas of cybersecurity and risk assessment in Cloud computing, autonomous vehicles, and Internet of Things.

SANJAY MADRIA

Sanjay Madria is Curators' Distinguished professor in the Department of Computer Science at the Missouri University of Science and Technology (formerly, University of Missouri-Rolla, USA). He received his Ph.D. in Computer Science from Indian Institute of Technology, Delhi, India in 1995. He has published over 250 journal and conference papers in the areas of mobile and sensor computing, mobile data management, cloud, and cyber security. He won five IEEE best papers awards in conferences such as IEEE MDM 2011, IEEE MDM 2012, and IEEE SRDS 2015. He is a co-author of a book published by Springer in 2003 and co-author of a new book by Morgan & Claypool in 2018. He is an Associate Editor of *Pervasive and Mobile Computing, Distributed and Parallel Databases*, and *Knowledge and Advanced Information System*. He has served/serving in international conferences as a general co-chair, pc co-chair, and steering committee member (such as of IEEE MDM, IEEE SRDS), and presented tutorials in the areas of secure sensor cloud, mobile computing, cloud computing, big data, etc. NSF, NIST, ARL,

ARO, AFRL, DOE, Boeing, and Hangsoft, among others, have funded his research projects. He has been awarded the JSPS (Japanese Society for Promotion of Science) visiting scientist fellowship in 2006 and the ASEE (American Society of Engineering Education) fellowship from 2008–2018. He was awarded the NRC Fellowship by National Academies in the year 2012 and 2018. He received faculty excellence research awards six times from his university for his research contributions. He is an ACM Distinguished Scientist and ACM Distinguished Speaker. He served as IEEE Distinguished Speaker and is an IEEE Senior Member as well as IEEE Golden Core Awardee.

Printed in the United States
by Baker & Taylor Publisher Services